Engineering Science
Second Level

Hutchinson TECtexts

Learning by Objectives
A Teachers' Guide
A. D. Carroll, J. E. Duggan & R. Etchells

Engineering Drawing and Communication
First Level
P. Collier & R. Wilson

Physical Science
First Level
A. D. Carroll, J. E. Duggan & R. Etchells

Electronics
Second Level
G. Billups & M. T. Sampson

Mathematics
Second Level
G. W. Allan & A. Hill

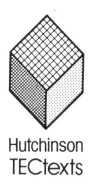

Hutchinson
TECtexts

Engineering Science

Second Level

D. Tipler, A. D. Carroll
and R. Etchells

Hutchinson of London

Hutchinson & Co. (Publishers) Ltd
3 Fitzroy Square, London W1P 6JD

London Melbourne Sydney Auckland
Wellington Johannesburg and agencies
throughout the world

First published 1979

© D. Tipler, A. D. Carroll and R. Etchells 1979
Illustrations © Hutchinson & Co. (Publishers) Ltd

Set in IBM Press Roman by Preface Ltd, Salisbury

Printed in Great Britain by The Anchor Press Ltd
and bound by Wm Brendon & Son Ltd,
both of Tiptree, Essex

ISBN 0 09 138371 4

Contents

Introduction

In each of the books in this series the authors have written text material to specified objectives. Test questions are provided to enable the reader to evaluate the objectives. The solutions or answers are given to all questions.

Topic area: Electric circuits and measurements

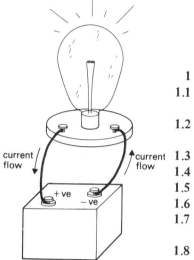

Figure 1 *Electron theory current flow*

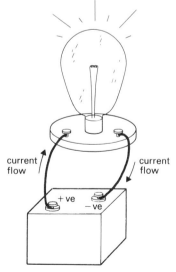

Figure 2 *Conventional current flow*

Section 1   Series-parallel circuits

After reading the following material, the reader shall:

1   Analyse simple series-parallel circuits.

1.1   State that in conventional current flow the current flows from the positive terminal to the negative terminal.

1.2   State that in electron theory current flow the current flows from the negative terminal to the positive terminal.

1.3   Identify conventional current flow.

1.4   Identify electron theory current flow.

1.5   Define current as the rate of movement of charge.

1.6   Define the ampere as 1 coulomb per second.

1.7   Define electromotive force in terms of total energy per coulomb produced.

1.8   Define potential difference in terms of energy per coulomb.

1.9   Define terminal potential difference as the voltage between the source terminals (and across the external circuit) when a current is flowing.

1.10 Distinguish between e.m.f., p.d. and terminal potential difference.

The reader will already have studied the concept of electric current flow; electrons in the outer orbits of most atoms can be quite easily forced out of their orbits and thus become free electrons. An electric current is the flow of these free electrons all moving in the same direction.

In the *electron theory of current flow* the electron drift is from negative to positive, that is, *the electric current flows from negative to positive*, as shown in Figure 1.

However, in *conventional current flow the electric current is assumed to flow from positive to negative*, as shown in Figure 2.

Note that conventional current flow will be used throughout the text.

In order to maintain a current flow it is essential that:

(*a*)   the circuit is complete, and

(*b*)   a force exists which maintains a difference in the state of charge between the terminals.

A quantity of electricity is often referred to as a *charge*. Because the basic unit of electrical charge is the electron, a quantity of electricity can be expressed as a number of electron charges. The unit is the *coulomb* where

1 coulomb (C) = 6.29×10^{18} electrons

Since an electric current is the flow of free electrons and because the electron is the basic unit of charge, an electric current is defined as the rate of movement of charge; that is, the number of electrons (or charges) passing a fixed point in a given time. Thus the fundamental unit of electric current is the *coulomb per second*. This unit is important, very frequently used, and is given the name *ampere*. When a charge of 1 coulomb passes a given point in a circuit in one second, an electric current (I) of 1 ampere (A) is flowing in the circuit.

1 ampere = 1 coulomb per second

As stated earlier, in order to maintain a current flow it is essential that the circuit contains some source, such as a battery, which will provide the force necessary to move the electrons round the circuit. This force is called the *electromotive force (e.m.f.)*. The symbol for e.m.f. is E, and the unit is the volt (V), so that 1 volt = 1 joule per coulomb.

Because the joule is a unit of energy, the e.m.f. expressed in volts is a measure of the energy available per unit of charge, i.e. the energy available per coulomb. Consider, for example, a 12 V battery used as an energy source. This means that an e.m.f. of 12 V is available for transfer to each coulomb.

Figure 3 shows a battery across which is connected in parallel a voltmeter (1). The battery is connected by a switch (4) to two resistors R_1 and R_2 which are connected in series. Connected in parallel with each resistor is a voltmeter (2) (3). In this theoretical circuit it is assumed that no energy is consumed by the instruments or the connecting wires.

When the switch (4) is open, no current flows in the circuit. The voltmeter (1) indicates the number of joules per coulomb which are available for transfer to the circuit from the source of electrical energy.

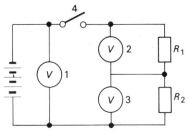

Figure 3   *Simple circuit*

When the switch (4) is closed, current moves round the circuit and energy is transferred from the current to the resistors. The voltmeters (2) and (3) indicate the *potential difference* in volts (joules/coulomb) across each resistor. The readings on the voltmeters are *potential drops*, and are often referred to as *volt drops*. Also, with the switch (4) closed, the voltmeter (1) indicates the *terminal potential difference*. The terminal potential difference is *the voltage between the source terminals (and across the external circuit) when a current is flowing*. Note that the terminal potential difference and the e.m.f. are not the same values. The terminal potential difference is smaller than the e.m.f. The reasons for this difference will be discussed later.

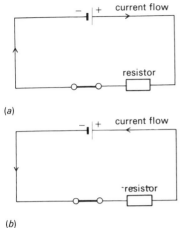

(a)

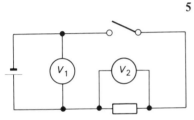

(b)

Figure 4   *Directions of current flow*

Figure 5   *e.m.f. – terminal voltage*

Self-assessment questions

Complete the following statements:

1   In conventional current flow the current is assumed to flow from the _____ terminal to the _____ terminal.

2   In electron current flow the current is assumed to flow from the _____ terminal to the _____ terminal.

3   Conventional electric current flows in the opposite direction to the flow of _____ .

4   Study Figure 4 (*a*) and (*b*) and then complete the following statements by crossing our the words that are incorrect.
Figure 4 (*a*) shows the current flowing according to CONVENTIONAL/ELECTRON theory current flow.
Figure 4 (*b*) shows the current flowing according to CONVENTIONAL/ELECTRON theory current flow.

5   Study Figure 5 and then complete the following statements:
(*a*)   The voltage measured across the terminals of the battery by V_1 when the switch is open is called the _____ .
(*b*)   The voltage measured across the terminals of the battery by V_1 when the switch is closed is called the _____ .
(*c*)   The voltage measured across the resistor by V_2 when the switch is closed is called the _____ .
For each of the following statements indicate whether they are true or false.

6   The p.d. between the terminals of any current carrying conductor is called voltage potential drop.
TRUE/FALSE

7   Current may be defined as the rate of movement of charge.
TRUE/FALSE

8   The total energy per coulomb produced by a battery is called the electromotive force.
TRUE/FALSE

9   An ampere of current is defined as 1 coulomb second.
TRUE/FALSE

10   The terminal p.d. is the p.d. which exists between the terminals of a battery when it is delivering current.
TRUE/FALSE

11   When a current is flowing in a circuit the value of the e.m.f. is the same as the terminal p.d.
TRUE/FALSE

12   The p.d. across the terminals of an electrical source is its potential drop.
TRUE/FALSE

13   The energy/coulomb required to cause a current to flow between two points in a material is called the e.m.f.

TRUE/FALSE

After reading the following material, the reader shall:

1.11   Define equivalent resistance as the single resistance which can replace a number of resistors.

1.12   Calculate the equivalent resistance for resistors in series.

1.13   Sketch series circuit diagrams from written information.

1.14   State the characteristics of a simple series circuit.

1.15   Solve problems involving simple series circuits.

In previous study on electricity the reader will have used the term *electrical resistance (R)*. Resistance is the term used to describe the ability of a material to impede the flow of electrons. The unit of resistance is the *ohm (Ω)*. When a resistor of any kind is inserted into a circuit the rate of electron flow is slowed down throughout the entire circuit. This means that the value of the current flowing is reduced, because the current is defined as the rate of movement of charge (i.e. electrons).

Different materials oppose the passage of electrons by different amounts. The magnitude of the opposition to the flow of electrons in a

---

Solutions to self-assessment questions

1   Positive to the negative terminal.

2   Negative to the positive terminal.

3   Electrons.

4   (*a*)   Current flowing negative to positive, i.e. electron theory.
     (*b*)   Current flowing positive to negative, i.e. conventional flow.

5   (*a*)   e.m.f.
     (*b*)   terminal p.d.
     (*c*)   potential drop.

6   TRUE.

7   TRUE.

8   TRUE.

9   FALSE. Current is rate of movement of charge, i.e. coulomb per second.

10   TRUE.

11   FALSE. The terminal p.d. is less than the e.m.f.

12   FALSE. The p.d. across the terminals of electrical source is either the e.m.f. if no current is flowing, or the terminal p.d. if a current is flowing.

13   TRUE.

material depends upon the atomic structure of the material. The atomic structure of materials such as glass, porcelain, p.v.c., polystyrene, rubber, wood and paper strongly oppose the passage of electrons. On the other hand materials such as copper, brass, aluminium, platinum, gold and silver offer very little opposition to the flow of electrons.

Electrical resistance is measured in volts per ampere, where

1 ohm (1 Ω) = 1 volt per ampere

That is, 1 ohm is the resistance which allows 1 ampere of current to flow when a p.d. of 1 volt is present across the resistance.

George Simon Ohm, a German physicist, suggested a law which provides a very important relationship between potential difference (V), resistance (R), and current (I). The relationship, known as Ohm's law, states that:

The potential difference between the ends of a conductor is directly proportional to the current flowing in it, provided that its temperature and other physical conditions do not change.

Ohm's law can be expressed by the equation

$V = IR$

where, V is expressed in volts
I is expressed in amperes
and R is expressed in ohms.

Example 1
A car headlamp bulb takes a current of 3 A from a 12 V battery. Calculate the resistance of the bulb filament.

Using $V = IR$ and transposing for R

$$R = \frac{V}{I}$$

$$V = 12 \text{ V and } I = 3 \text{ A}$$

$$\therefore R = \frac{12}{3} \frac{[\text{volts}]}{[\text{amperes}]} = 4 \Omega$$

Resistance of lamp filament = 4 Ω.

Example 2
A current of 250 mA flows through a resistance of 500 Ω. Calculate the p.d. across the 500 Ω resistance.

Expressing the current in amperes:

$$250 \text{ mA} = \frac{250}{1000} \text{ A} = 0.25 \text{ A}$$

Using $V = IR$
when $I = 0.25$ A and $R = 500\ \Omega$
$$\therefore V = 0.25\ [\text{ampere}] \times 500\ [\text{ohm}]$$
$$\therefore V = 125\ \text{V}$$
Potential difference across resistance is 125 V.

Self-assessment questions

14 Calculate the potential difference across a length of wire having a resistance of 20 Ω when a current of 250 mA is flowing through it.

15 If a voltage of 60 kV is measured across a resistance of 4 MΩ what is the value of the current flowing?

Resistors can be connected to sources of e.m.f. in three ways:

(i)   series
(ii)   parallel
(iii) a combination of series and parallel.

Each of these three methods of connecting resistors will be considered in detail. The term *equivalent resistance* will be used in all three cases. The equivalent resistance is the single resistance that can replace a number of resistors. The use of this single resistance makes easier the analysis of more complex electrical circuits.

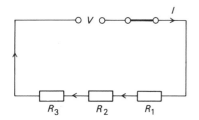

Figure 6   *Resistors in series*

A series circuit is formed when two or more resistors are connected end to end in a circuit, in such a way that there is only one path for current flow, as shown in Figure 6.

The characteristics of a simple series circuit such as shown in Figure 7 are as follows:

1 The value of the current flowing through a series circuit is always the same at every point in the circuit.

2 The total resistance in a series circuit is always the sum of the individual values of resistance in the circuit. That is:

$$R_{\text{total}} = R_1 + R_2 + R_3 + \text{ etc.}$$

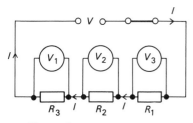

Figure 7   *Simple series circuit*

3 For the series circuit the equivalent resistance R is the sum of the individual resistances. That is:

$$R = R_1 + R_2 + R_3 + \text{ etc.}$$

4 The voltage drop across the circuit resistance is always equal to the sum of the voltage drops across the individual resistances. That is:

$$V = V_1 + V_2 + V_3 + \text{ etc.}$$

Example 3

If a current of 1.5 A flows through two resistances of 8 Ω and 4 Ω connected in series, calculate:

(a)   the p.d. across each resistance

(b)   the supply voltage.

Figure 8 shows the circuit in which $I = 1.5$ A.

(a)   Let p.d. across R_1 be V_1.

Let p.d. across R_2 be V_2.

To find p.d. across R_1 apply Ohm's law,

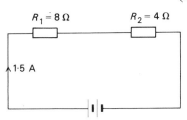

Figure 8   *Two resistors in series*

i.e. $V_1 = IR_1$

$\therefore \quad V_1 = 1.5$ [ampere] × 8 [ohms]

$\quad V_1 = 12$ V

To find p.d. across R_2 apply Ohm's law,

i.e. $V_2 = IR_2$

$\therefore \quad V_2 = 1.5$ [amperes] × 4 [ohms]

$\quad V_2 = 6$ V

(b)   To find the supply voltage V,

$V = V_1 + V_2$

$\therefore \quad V = 12 + 6 = 18$ V

Alternatively,

total resistance $= R = R_1 + R_2$

$\therefore R = 8 + 4 = 12$ Ω

Applying Ohm's law,

i.e. $V = IR$

$\therefore \quad V = 1.5 × 12 = 18$ V

Example 4

For the simple series circuit shown in Figure 9, the p.d. across R_1 is 5 V. Calculate the p.d. across R_2 and R_3 and the total supply voltage. Consider resistor R_1.

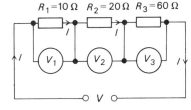

Figure 9   *Three resistors in series*

Applying Ohm's law, i.e. $V_1 = IR_1$

$$I = \frac{V_1}{R_1}$$

$$I = \frac{5}{10} = 0.5 \text{ A}$$

To find the p.d. across R_2.

Apply Ohm's law, i.e. $V_2 = IR_2$

$\therefore V_2 = 0.5 × 20 = 10$ V

To find the p.d. across R_3.

Apply Ohm's law, i.e. $V_3 = IR_3$

$\therefore V_3 = 0.5 × 60 = 30$ V

Let the total supply voltage = V.

Now $V = V_1 + V_2 + V_3$
     $\therefore V = 5 + 10 + 30$
     $\therefore V = 45$ V

Alternatively, applying Ohm's law, i.e. $V = IR$,
where $R = R_1 + R_2 + R_3$
      $R = 10 + 20 + 60 = 90$ Ω
     $\therefore V = 0.5 \times 90$ V
       $V = 45$ V

Self-assessment questions

16   The equivalent resistance is the single resistance which can be used to replace a number of resistors.

TRUE/FALSE

17   Complete the equation for equivalent resistance for resistors connected in series. $R = $ _____ .

18   If four resistors of value 4, 6, 8, and 10 Ω are connected in series, what is the equivalent resistance of the circuit?

Questions 19 to 23 are based on the circuit in Figure 10.

19   The resistors shown are connected in _____ .

20   The total resistance of the resistors is given by $R = $ _____ .

21   The resistors R_1 and R_2 may be replaced by their equivalent resistance without altering the current in the system.

TRUE/FALSE

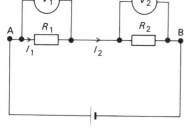

Figure 10   *Series circuit for self-assessment questions 19-23*

22   The value of the current I_1 passing through resistor R_1 is the same as the value of the current I_2 passing through resistor R_2.

TRUE/FALSE

23   The total p.d. between points A and B is given by $V = V_1 + V_2$.

TRUE/FALSE

Solutions to self-assessment questions

14    $V = IR$ where $I = 250 \times 10^{-3}$ A and $R = 20$ Ω
     $\therefore V = 250 \times 10^{-3} \times 20 = 5$ V.

15    $I = \dfrac{V}{R}$

where $V = 60 \times 10^3$ V and $R = 4 \times 10^6$ Ω

     $\therefore I = \dfrac{60 \times 10^3}{4 \times 10^6} = 0.015$ A

24   A current of 2 A flows through two resistances of 6 Ω and 4 Ω connected in series. The e.m.f. for the circuit is supplied by a battery.
Sketch the circuit and calculate the supply voltage and the p.d. across
each resistance.

After reading the following material, the reader shall:

1.16   Calculate the equivalent resistance for resistors in parallel.
1.17   Sketch parallel circuit diagrams from written information.
1.18   State the characteristics of simple parallel circuits.
1.19   Solve problems involving simple parallel circuits.

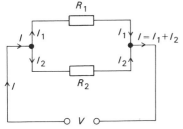

Figure 11   *Simple parallel circuit*

When resistors, instead of being connected end to end as in a series
circuit, are connected side by side so that there exists more than one
path through which current can flow, the resistors are said to be connected in parallel, and the circuit of which they form a part is called a
parallel circuit.

The characteristics of a simple parallel circuit such as that shown in
Figure 11 are as follows:

1   The current divides to flow through the parallel branches of the circuit,
so that $I = I_1 + I_2 +$ etc.

2   The p.d. across every resistance in a parallel circuit is the same, and is
equal to that of the voltage source. So that applying Ohm's law to the
individual branch resistances:

$$I_1 = \frac{V}{R_1}$$

$$I_2 = \frac{V}{R_2}$$

Adding the two equations above to get I:

$$I = I_1 + I_2 = \frac{V}{R_1} + \frac{V}{R_2}$$

Now consider a single equivalent resistance R which is the exact equivalent of the two in parallel. The p.d. across this resistance R is V when it
is carrying the total current I.

Applying Ohm's law to this equivalent resistance R,

$$I = \frac{V}{R}$$

But $I = I_1 + I_2 = \dfrac{V}{R}$

Replacing $I_1 + I_2$ with $\dfrac{V}{R_1} + \dfrac{V}{R_2}$,

$$\frac{V}{R_1} + \frac{V}{R_2} = \frac{V}{R}$$

Dividing through by V which is common, the equation becomes,

$$\frac{1}{R_1} = \frac{1}{R_1} + \frac{1}{R_2}$$

If the circuit contains more than two resistors in parallel, then the equation becomes,

$$\frac{1}{R} = \frac{1}{R_1} + \frac{1}{R_2} + \frac{1}{R_3} + \text{etc.}$$

That is, for resistances in parallel, the reciprocal of the equivalent resistance equals the sum of the reciprocals of the individual resistances. Note that in a parallel circuit the value of the total resistance is lower than the value of the smallest individual resistance in the circuit.

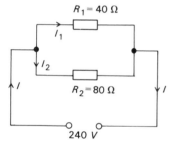

Figure 12   *Equivalent resistance*

Example 5
For the parallel circuit shown in Figure 12, calculate:
(a)   the equivalent resistance,
(b)   the total circuit current and
(c)   the current through each resistor.

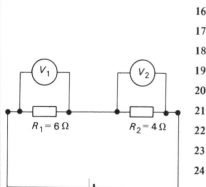

Figure 13   *Solution to self-assessment question 24*

Solutions to self-assessment questions

16   TRUE.

17   $R = R_1 + R_2 + R_3 + \text{etc.}$

18   $R = 4 + 6 + 8 + 10 = 28$

19   Series.

20   $R = R_1 + R_2$

21   TRUE.

22   TRUE.

23   TRUE.

24   Equivalent resistance $= R_1 + R_2 = 6 + 4 = 10\ \Omega$
Supply voltage $= V = IR = 2 \times 10 = 20$ V
p.d. across 6 Ω resistor: $V_1 = IR_1 = 2 \times 6 = 12$ V
p.d. across 4 Ω resistor: $V_2 = IR_2 = 2 \times 4 = 8$ V
The circuit should be similar to that shown in Figure 13.

(*a*)   Let the equivalent resistance be R.

Then $\dfrac{1}{R} = \dfrac{1}{R_1} + \dfrac{1}{R_2}$

$\therefore \quad \dfrac{1}{R} = \dfrac{1}{40} + \dfrac{1}{80}$

$\therefore \quad \dfrac{1}{R} = \dfrac{2+1}{80} = \dfrac{3}{80}$

$\therefore \quad R = \dfrac{80}{3} = 26.67 \ \Omega$

(*b*)   Using Ohm's law, $V = IR$.

Transpose to give $I = \dfrac{V}{R}$

$\therefore \quad I = \dfrac{240}{26.67} = 9 \text{ A}$

(*c*)   To find the current through R_1

$I_1 = \dfrac{V}{R_1} = \dfrac{240}{40}$

$\therefore \ I_1 = 6 \text{ A}$

To find the current through R_2

$I_2 = \dfrac{V}{R_2} = \dfrac{240}{80}$

$\therefore \ I_2 = 3 \text{ A}$

Note that $I = I_1 + I_2 = 9 \text{ A}$

Example 6

For the parallel circuit shown in Figure 14 calculate the value of the resistor R_2.

Let R = equivalent resistance.
Applying Ohm's law,

$R = \dfrac{V}{I}$

$R = \dfrac{6}{1} = 6 \ \Omega$

Now for resistances in parallel,

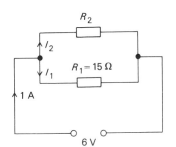

Figure 14   *Resistors in parallel*

$$\frac{1}{R} = \frac{1}{R_1} + \frac{1}{R_2}$$

$$\therefore \frac{1}{6} = \frac{1}{15} + \frac{1}{R_2}$$

$$\therefore \frac{1}{R_2} = \frac{1}{6} - \frac{1}{15}$$

$$\therefore \frac{1}{R_2} = \frac{5-2}{30} = \frac{3}{30}$$

$$\therefore R_2 = \frac{30}{3} = 10\ \Omega$$

The value of the resistor R_2 is 10 Ω.

Self-assessment questions

Questions 25 to 30 relate to the circuit shown in Figure 15.

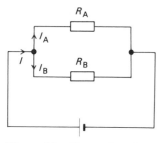

25   The resistors connected as shown in the circuit are connected in _____ .

26   The p.d. across R_A is different from the p.d. across R_B.
                                                    TRUE/FALSE

27   The general formula for resistors connected in this manner is given by $1/R =$

28   The resistors R_A and R_B may be replaced by an equivalent resistor without altering the value of the current I.
                                                    TRUE/FALSE

Figure 15   *Circuit for self-assessment questions 25-30*

29   The equivalent resistance R for the circuit is given by the expression $R = R_A + R_B$.
                                                    TRUE/FALSE

30   The relationship between I, I_A and I_B is $I = I_A + I_B$.
                                                    TRUE/FALSE

31   For each of the following sets of resistors connected in parallel calculate the value of the single equivalent resistance.
      (*a*)   20 Ω, 40 Ω
      (*b*)   20 Ω, 40 Ω, 80 Ω
      (*c*)   10 Ω, 20 Ω, 30 Ω, 40 Ω.

32   Two resistors of 10 Ω and 30 Ω are connected in parallel across a 240 V supply.
      (*a*)   sketch the circuit
      (*b*)   calculate the equivalent resistance of the circuit
      (*c*)   calculate the value of the current flowing in each branch of the circuit.

33   Three resistors R_1, R_2 and R_3 have values of 3, 9 and 12 Ω respectively and are connected in parallel. The value of the current flowing through the 9 Ω resistor is 0.5 A. Sketch the circuit described above and calculate the value of the supply p.d. and the circuit current.

After reading the following material, the reader shall:

1.20   Calculate the equivalent resistance for a series-parallel combination of resistors.

1.21   Sketch the current path through a series-parallel circuit.

1.22   Calculate the values of currents and potential differences for series-parallel combinations.

1.23   Solve problems involving simple series-parallel arrangements of resistors supplied from a single d.c. source.

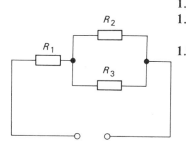
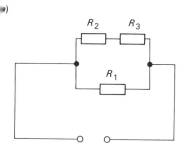

Figure 16   *Series-parallel circuits*

Circuits consisting of three or more resistors may be connected to form a series-parallel circuit. Figure 16 illustrates the two basic types of series-parallel circuits. Figure 16 (*a*) shows a resistor connected in series with a parallel combination, and Figure 16 (*b*) shows a circuit where one or more of the branches of a parallel circuit consist of resistors in series. Such combinations of resistors are frequently used in electrical circuits, particularly in electric motors and in control circuits for electrical equipment.

The basic steps in finding the single equivalent resistance of a series-parallel circuit are as follows:

1   Redraw the circuit if necessary.

2   If any of the parallel parts of the circuit have branches consisting of two or more resistors in series, find the total value of these resistors by adding them, using the equation $R = R_1 + R_2 + R_3 +$ etc.

3   Using the formula for parallel resistances,

i.e. $\dfrac{1}{R} = \dfrac{1}{R_1} + \dfrac{1}{R_2} + \dfrac{1}{R_3} +$ etc.,

find the total resistance of the parallel parts of the circuit.

4   Add the combined parallel resistances to any resistances which are in series with them.

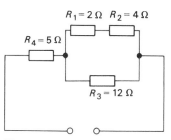

Figure 17   *Resistors in series-parallel*

Example 7

Find the single equivalent resistance for the series-parallel circuit shown in Figure 17.

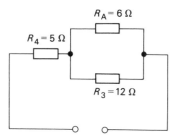

Figure 18   *Equivalent resistance of series branch resistors*

First combine the two series branch resistors R_1 and R_2 by adding them to give an equivalent resistance R_A, and to give the circuit shown in Figure 18:

$$R_A = R_1 + R_2$$
$$R_A = 2 + 4 = 6\ \Omega$$

Next the parallel combination of R_3 and R_A is combined using the parallel resistance formula, to give an equivalent resistance R_B, and to give the circuit shown in Figure 21:

Solutions to self-assessment questions

25   Parallel.

26   FALSE.

27 $\dfrac{1}{R} = \dfrac{1}{R_1} + \dfrac{1}{R_2} + \dfrac{1}{R_3}$ + etc.

28   TRUE.

29   FALSE; the p.d. across the two resistors in parallel is the same.

30   TRUE.

31 $(a)\ \dfrac{1}{R} = \dfrac{1}{20} + \dfrac{1}{40} = \dfrac{2+1}{40}$ $\therefore R = \dfrac{40}{3} = 13.33\ \Omega$

$(b)\ \dfrac{1}{R} = \dfrac{1}{20} + \dfrac{1}{40} + \dfrac{1}{80} = \dfrac{4+2+1}{80}$ $\therefore R = \dfrac{80}{7} = 11.43\ \Omega$

$(c)\ \dfrac{1}{R} = \dfrac{1}{10} + \dfrac{1}{20} + \dfrac{1}{30} + \dfrac{1}{40} = \dfrac{12+6+4+3}{120}$ $\therefore R = \dfrac{120}{25} = 4.8\ \Omega$

32 $(b)\ \dfrac{1}{R} = \dfrac{1}{10} + \dfrac{1}{30} = \dfrac{3+1}{30}$

$\therefore$ equivalent resistance $= \dfrac{30}{4} = 7.5\ \Omega$

$(c)\ I_1 = \dfrac{240}{10} = 24\ \text{A}$

$I_2 = \dfrac{240}{30} = 8\ \text{A}$

33   For R_2, $V = I_2 R_2 = 0.5 \times 9 = 4.5\ \text{V} = $ supply voltage

$I_1 = \dfrac{V}{R_1} = \dfrac{4.5}{3} = 1.5\ \text{A}$

$I_3 = \dfrac{V}{R_3} = \dfrac{4.5}{12} = 0.375\ \text{A}$

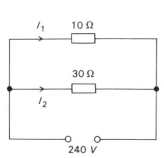

Figure 19   *Solution to self-assessment question 32*

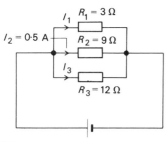

Figure 20   *Solution to self-assessment question 33*

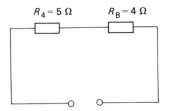

Figure 21 *Equivalent series circuit*

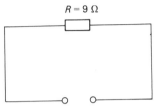

Figure 22 *Single equivalent resistance*

$$\frac{1}{R_B} = \frac{1}{R_3} + \frac{1}{R_A}$$

$$\frac{1}{R_B} = \frac{1}{12} + \frac{1}{6}$$

$$\frac{1}{R_B} = \frac{1+2}{12} = \frac{3}{12}$$

$$\therefore R_B = \frac{12}{3} = 4\,\Omega$$

Finally, the series resistor R_4 is added to the equivalent resistance R_B. To give the single equivalent resistance R for the whole circuit, as shown in Figure 22:

$$R = R_4 + R_B$$
$$R = (5+4)\,\Omega$$
$$\therefore \quad R = 9\,\Omega$$

More complicated circuits only require more stages of calculation. They do not require more formulae.

Example 8
Calculate the equivalent resistance for the series-parallel circuit shown in Figure 23:

First combine the two series resistors R_1 and R_2 to given an equivalent resistance R_A.

$$R_A = R_1 + R_2$$
$$R_A = 2+4 = 6\,\Omega$$

Now combine the two series resistors R_3 and R_4 to give an equivalent resistance R_B, and a circuit as shown in Figure 24:

$$R_B = R_3 + R_4$$
$$R_B = 3+9 = 12\,\Omega$$

Now, for the two parallel resistors R_A and R_B calculate a single equivalent resistance R_C, to give a circuit as shown in Figure 25.

$$\frac{1}{R_C} = \frac{1}{R_A} + \frac{1}{R_B}$$

$$\frac{1}{R_C} = \frac{1}{6} + \frac{1}{12}$$

$$\frac{1}{R_C} = \frac{2+1}{12} = \frac{3}{12}$$

$$\therefore \quad R_C = 4\,\Omega$$

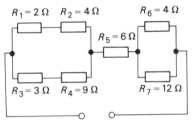

Figure 23 *More complex series-parallel circuit*

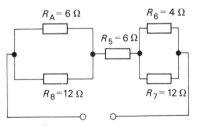

Figure 24 *Series branch combined*

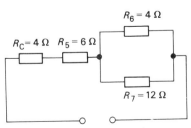

Figure 25 *Equivalent series resistance*

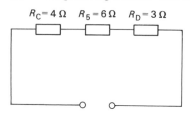

Figure 26   *Equivalent resistors in series*

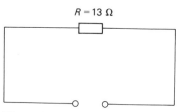

Figure 27   *Equivalent single resistance*

For the two parallel resistors R_6 and R_7 calculate a single equivalent resistance R_D.

$$\frac{1}{R_D} = \frac{1}{R_6} + \frac{1}{R_7}$$

$$\frac{1}{R_D} = \frac{1}{4} + \frac{1}{12}$$

$$\frac{1}{R_D} = \frac{3+1}{12} = \frac{4}{12}$$

$$\therefore \quad R_D = 3\ \Omega$$

Finally add the three series resistors R_C, R_5 and R_D to give a single equivalent resistance R for the series-parallel circuit shown in Figure 23.

$$R = R_C + R_5 + R_D$$
$$R = 4 + 6 + 3$$
$$R = 13\ \Omega$$

Self-assessment question

34   For each of the series-parallel circuits shown in Figures 28 to 31 inclusive, calculate the single equivalent resistance.

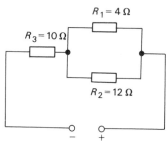

Figure 28

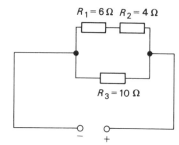

Figure 29

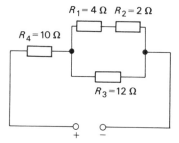

Figure 30

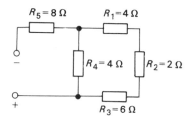

Figure 31

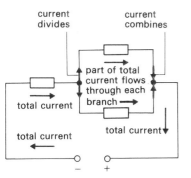

Figure 32   *Current flowing in a series-parallel circuit*

The total circuit current for a series-parallel circuit depends on the total resistance offered by the circuit when connected across a voltage source. Figure 32 illustrates how the current flows through a series-parallel circuit. Current in the circuit divides to flow through all parallel paths, and comes together again to flow through series parts of the circuit.

Voltage drops across a series-parallel circuit occur in the same way as they do in series and parallel circuits. The voltage drops across the resistors in the series part of the circuit depend on the individual values of the resistors. In the parallel parts of the circuit, every resistor has the same p.d. across it. Figure 33 illustrates how the p.d. is distributed in a series-parallel circuit. Note that in the parallel part of the circuit the voltage drop across each branch is the same, that is, $V_B = V_F = 50$ V.

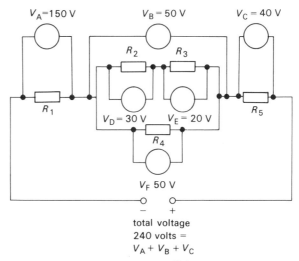

Figure 33   *Voltage drops in series-parallel circuit*

Note also, that in the branch of the parallel part of the circuit that has the two resistors R_2 and R_3, the sum of the voltage drops across these two resistors is equal to the total voltage drop across the two resistors, that is, $V_D + V_E = V_B$.

Example 9
Figure 35 shows a series-parallel circuit connected across a 20 V supply. Calculate

(a)   the equivalent resistance,
(b)   the total circuit current,
(c)   the voltmeter readings on V_1 and V_2,
(d)   the ammeter reading on A_1.

(*a*)   Let the equivalent resistance be R Ω. Reduce the three parallel resistors to a single equivalent resistance R_A.

where $\dfrac{1}{R_A} = \dfrac{1}{4} + \dfrac{1}{10} + \dfrac{1}{20}$

$\therefore$ $\dfrac{1}{R_A} = \dfrac{5 + 2 + 1}{20} = \dfrac{8}{20}$

$\therefore$ $R_A = \dfrac{20}{8} = 2.5\ \Omega$

Solution to self-assessment question

34(i)   For parallel resistors

$$\frac{1}{R} = \frac{1}{4} + \frac{1}{12} = \frac{3 + 1}{12}$$

$\therefore$ $R = \dfrac{12}{4} = 3\ \Omega$ (see figure 28)

For 10 Ω and 3 Ω in series equivalent resistance $R = R_3 + 3 = 10 + 3 = 13\ \Omega$.

(ii)   For the resistors in series, equivalent resistance is $6 + 4 = 10\ \Omega$ (see Figure 29). For the 10 Ω and R_3 in parallel, the equivalent resistance R is given by,

$$\frac{1}{R} = \frac{1}{10} + \frac{1}{10} = \frac{2}{10}$$

$\therefore$ $R = \dfrac{10}{2} = 5\ \Omega$

(iii)   For the resistors in series on the parallel section of the circuit, the equivalent resistance is $4 + 2 = 6\ \Omega$ (see Figure 30). For this 6 Ω in parallel with the 12 Ω resistor the equivalent resistance is given by,

$$\frac{1}{R} = \frac{1}{12} + \frac{1}{6} = \frac{1 + 2}{12} = \frac{3}{12}$$

$\therefore$ $R = \dfrac{12}{3} = 4\ \Omega$

For the 4 Ω and R_4 in series, the equivalent resistance is $R = 4 + 10 = 14\ \Omega$.

(iv)   Part of Figure 31 is redrawn as shown in Figure 34. R_1, R_2 and R_3 in series can be replaced by one resistor of value $R_1 + R_2 + R_3 = 12\ \Omega$. This 12 Ω resistance is now in parallel with R_4. The equivalent resistance is

$$\frac{1}{R} = \frac{1}{12} + \frac{1}{4} = \frac{1 + 3}{12} = \frac{4}{12}$$

$\therefore$ $R = \dfrac{12}{4} = 3\ \Omega$

This 3 Ω resistance is in series with the resistor R_5. The equivalent resistance is $3 + R_5 = (3 + 8) = 11\ \Omega$.

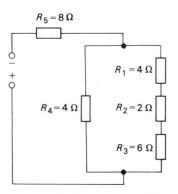

Figure 34 *Solution to self-assessment question 34*

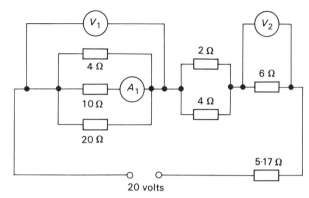

Figure 35   *Series-parallel circuit for Example 9*

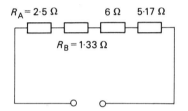

Figure 36   *Four resistors in series*

Reduce the two parallel resistors to a single equivalent resistance R_B:

where $\quad \dfrac{1}{R_B} = \dfrac{1}{2} + \dfrac{1}{4}$

$\therefore \quad \dfrac{1}{R_B} = \dfrac{2+1}{4} = \dfrac{3}{4}$

$\therefore \quad R_B = \dfrac{4}{3} = 1.33 \ \Omega$

The circuit can now be represented by four resistors in series as shown in Figure 36. The single equivalent resistance is,

$R = (R_A + R_B + 6 + 5.17) \ \Omega$
$R = (2.5 + 1.33 + 6 + 5.17) \ \Omega$
$R = 15 \ \Omega$

The single equivalent resistance is 15Ω.

(b)   To find the total circuit current use the relationship,

$$I = \frac{V}{R}$$

$\therefore \quad I = \dfrac{20}{15} = 1.25 \ \text{A}$

The total circuit current is 1.25 A.

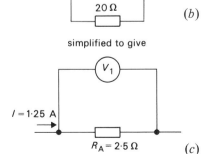

Figure 37   *Parallel branch reduced to single resistance*

(c)   To find the reading on voltmeter V_1 consider the three parallel resistor part of the circuit simplified as shown in Figure 37. To find the voltage drop V_1 use the relationship,

$$V = IR$$
i.e. $V_1 = IR_A$
$\therefore \quad V_1 = 1.25 \times 2.5$
$\therefore \quad V_1 = 3.125 \text{ V}$

To find the reading on voltmeter V_2, again use the relationship,

$$V = IR$$
$\therefore \quad V_2 = I \times 6$ (see Figure 35.)
$\therefore \quad V_2 = 1.25 \times 6$
$\therefore \quad V_2 = 7.5 \text{ V}$

(d)   To find the ammeter reading A_1 use the relationship,

$$I = \frac{V}{R}$$

i.e. $I_1 = \dfrac{V_1}{10}$ (see Figure 35.)

$\therefore \quad I_1 = \dfrac{7.5}{10} = 0.75 \text{ A}$

Self-assessment questions

Figure 38 and 39 show series-parallel circuits. Voltmeters denoted by Ⓥ and ammeters denoted by Ⓐ are shown at various points in the circuit.

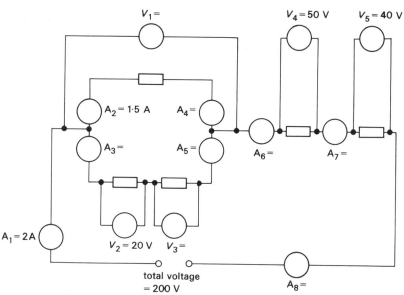

Figure 38   *Circuit for self-assessment question 35*

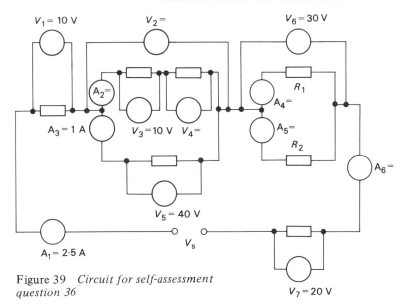

Figure 39   *Circuit for self-assessment question 36*

35   For the circuit shown in Figure 38 complete the readings for the voltmeters V_1 and V_3 and for the ammeters A_3, A_4, A_5, A_6, A_7 and A_8.

36   For the circuit shown in Figure 39,
(*a*) What are the readings on V_2 *and* V_4?
(*b*) What is the value of V_5?
(*c*) What are the readings on A_2 and A_6?
(*d*) If the values of R_1 and R_2 are the same, what are the readings on A_4 and A_5?

37   The series-parallel combination circuit shown in Figure 40 has a supply of 12 V. Determine,
(*a*) the equivalent resistance of the circuit ,
(*b*) the total circuit current,
(*c*) the current passing through each branch of the parallel section of the circuit,
(*d*) the voltage drop across the resistor R_5.
Draw a sketch to illustrate the path of the current round the circuit.

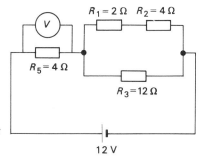

Figure 40   *Circuit for self-assessment question 37*

38   The series-parallel combination circuit shown in Figure 41 has a supply voltage of 20 V. Determine,

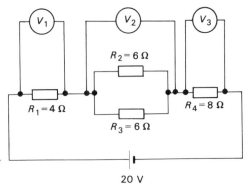

Figure 41   *Circuit for self-assessment question 38*

20 V

(*a*) the equivalent resistance of the circuit,
(*b*) the total circuit current,
(*c*) the voltmeter readings V_1, V_2 and V_3,
(*d*) the value of the current passing through resistor R_2.
        Draw a sketch to illustrate the path of the current round the circuit.

Solutions to self-assessment questions

35 V_1 = $200 - (V_4 + V_5)$ = 110 V.
           V_3 = $V_1 - V_2$ = 90 V.
           A_3 = $A_1 - A_2 = 0.5$ A.
           A_4 = $A_2 = 1.5$ A.
           A_5 = $A_3 = 0.5$ A.
           A_6 = $A_7 = A_8 = A_1 = 2$ A = total current.

36   (*a*) $V_2 =$ $V_5 = 40$ V.
                 $V_4 =$ $40 - 10 = 30$ V.
           (*b*) Circuit p.d. = $V_1 + V_2 + V_6 + V_7 = 10 + 40 + 30 + 20 = 100$ V.
                 $A_6 =$ total current = 2.5 A.
           (*c*) $A_2 = A_1 - A_3 = 2.5 - 1 = 1.5$ A.
           (*d*) $A_4 =$ $A_5 = 2.5/2 = 1.25$ A. Each resistor takes half the current.

37   (*a*) Equivalent resistance of circuit.
                 For parallel section. Combine the R_1 and R_2 to give 6 Ω.

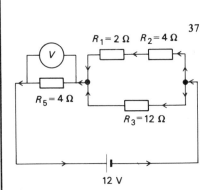

12 V
Figure 42   *Current path round circuit shown in Figure 40*

$$\text{Then } \frac{1}{R} = \frac{1}{6} + \frac{1}{12} = \frac{3}{12} . \ \therefore R = 4 \ \Omega.$$

                 Add this 4 to R_5 to get equivalent $R = 8$ Ω.
           (*b*) Total circuit current $I = \dfrac{V}{R} = \dfrac{12}{8} = 1.5$ A.

           (*c*) p.d. across parallel section = $IR = 1.5 \times 4 = 6$ V.
                 Therefore current through R_3 is

$$\frac{V}{R_3} = \frac{6}{R_3} = \frac{6}{12} = 0.5 \ \text{A.}$$

                 Therefore current through R_1 and R_2 is $1.5 - 0.5 = 1.0$ A.
           (*d*) Voltage drop across R_5 is $IR_5 = 1.5 \times 4 = 6$ V.
           See Figure 42 for current path round circuit.

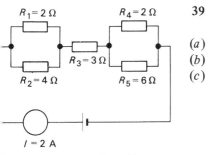

Figure 43 *Circuit for self-assessment question 39*

39 The series-parallel combination circuit shown in Figure 43 has a total circuit current of 2 A. Determine,

(*a*) the p.d. of the supply,
(*b*) the voltage drop across the parallel resistors R_1 and R_2,
(*c*) the current flowing through the resistor R_1.

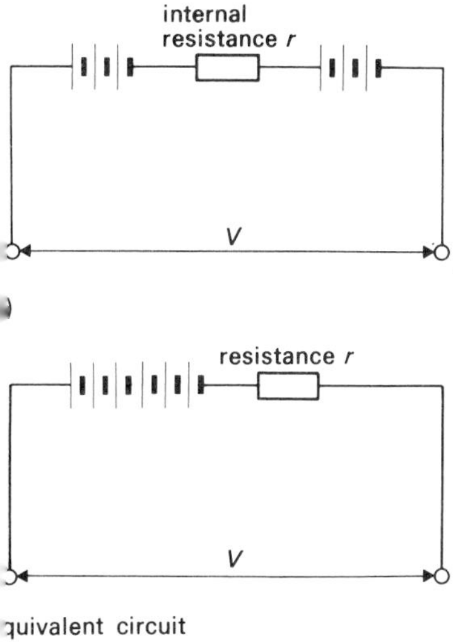

After reading the following material, the reader shall:

1.24 Define internal resistance as the resistance offered by the source to the flow of current.

1.25 State the equation for terminal voltage V as $V = E \times \dfrac{R}{R + r}$.

1.26 Solve problems involving internal resistance.

Figure 44 *Equivalent circuit for internal resistance*

In all the calculations to date it has been assumed that the electrical generator (i.e. the battery or some other source of e.m.f.) has been 100 per cent electrically efficient. This assumption is not true because any generator of electric current will have *internal resistance*. Internal resistance is defined as *the resistance offered by the source to the flow of current.*

In the conventional current flow the current is assumed to flow from the positive terminal to the negative terminal. Because the source has a resistance, there is a loss in potential in the source itself, and the effective p.d. across its terminals is not quite as high as it theoretically should be.

An equivalent circuit can be used to allow for the effect of internal resistance. Figure 44 (*b*) shows the simplified version of the complicated action which is taking place in Figure 44 (*a*).

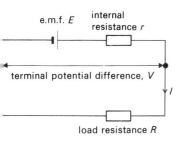

Figure 45 *Internal resistance and load resistance*

The p.d. across the source terminals — the terminal potential difference — is less than the full e.m.f. of the source. To obtain an expression for this terminal voltage, consider the circuit as shown in Figure 45.

In this circuit,

E represents the e.m.f. of the battery,
V represents the p.d. between the battery terminals, when the battery is supplying current,
r represents the internal resistance of the battery,
R represents the total external resistance or load resistance.

The total resistance for the circuit is $R + r$. The circuit current can be calculated from $I = \dfrac{E}{R + r}$ (from $I = \dfrac{V}{R}$). However the true p.d. between the terminals is the value that can actually be measured. Now for the circuit shown $V = IR$. Substituting for $I = \dfrac{E}{R + r}$ gives

$$V = E \times \frac{R}{R + r}$$

Solutions to self-assessment questions

38   (a)   For parallel part, $\dfrac{1}{R} = \dfrac{1}{6} + \dfrac{1}{6} = \dfrac{2}{6}$. $\therefore R = 3\ \Omega$.

Therefore equivalent resistance is $R_1 + 3 + R_4 = 15\ \Omega$.

(b)   Total circuit current $I = \dfrac{V}{R} = \dfrac{20}{15} = 1.33$ A.

(c) $V_1 = IR_1 = 1.33 \times 4 = 5.32$ V.
$V_2 = IR = 1.33 \times 3 = 3.99$ V.
$V_3 = IR_4 = 1.33 \times 8 = 10.64$ V.

(d)   Current through R_2 is $\dfrac{1.33}{2} = 0.665$ A. Since the two resistors have the same values they each take half the current.

See Figure 46 for current path round circuit.

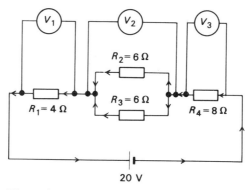

20 V

Figure 46   *Current path round circuit shown in Figure 41*

39   (a)   For R_1 and R_2, $\dfrac{1}{R} = \dfrac{1}{2} + \dfrac{1}{4} = \dfrac{3}{4}$. $\therefore R = \dfrac{4}{3} = 1.33\ \Omega$.

For R_4 and R_5, $\dfrac{1}{R} = \dfrac{1}{2} + \dfrac{1}{6} = \dfrac{4}{6}$. $\therefore R = \dfrac{6}{4} = 1.5\ \Omega$.

Therefore equivalent $R = 1.33 + 1.5 + 3 = 5.83\ \Omega$.
Hence supply p.d. is $IR = 2 \times 5.83 = 11.66$ V.

(b)   Voltage drop across R_1 and R_2 is $IR = 2 \times 1.33 = 2.66$ V.

(c)   Current through R_1 is $\dfrac{V}{R} = \dfrac{2.66}{2} = 1.33$ A.

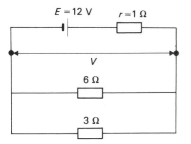

Figure 47 *Circuit for Example 10*

Example 10

The circuit shown in Figure 47 represents a battery of e.m.f. 12 V and internal resistance 1 Ω, connected in parallel with two resistors of 6 Ω and 3 Ω.

Calculate (*a*) the terminal voltage V,

(*b*) the circuit current I,

(*c*) the voltage drop in the battery due to the internal resistance.

(*a*) The terminal voltage V can be calculated from the equation

$$V = E \times \frac{R}{R + r}$$

where R = total external resistance or load resistance. To obtain R, reduce the two resistances in parallel to a single equivalent resistance, R_E. That is,

$$\frac{1}{R_E} = \frac{1}{6} + \frac{1}{3}$$

$$\therefore \frac{1}{R_E} = \frac{1 + 2}{6} = \frac{3}{6}$$

$$\therefore R_E = \frac{6}{3} = 2 \ \Omega$$

The circuit can now be reduced to a simple series circuit as shown in Figure 48.

$$\text{Now } V = E \times \frac{R_E}{R_E + r}$$

$$\therefore V = 12 \times \frac{2}{2 + 1}$$

$$\therefore V = 12 \times \frac{2}{3} = 8 \text{ V}$$

The terminal voltage is 8 V.

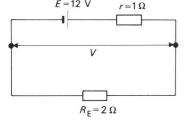

Figure 48 *Equivalent series circuit for Figure 47*

(*b*) The circuit current $I = \dfrac{V}{R_E}$

$$\therefore I = \frac{8}{2} = 4 \text{ A}$$

(*c*) The voltage drop in the battery can be calculated by considering the difference between the values of E and V,

that is, voltage drop = $12 - 8 = 4$ V

Alternatively, by considering the internal resistance, the voltage drop across this resistance is given by,

$$V = Ir$$

That is $V = 4 \times 1 = 4\,V$

What are the practical consequences of internal resistance? Consider the equation $V = E \times \dfrac{R}{R + r}$. If the load resistance is high and the internal resistance is low, then the terminal voltage V approximates very closely to the potential e.m.f. For example, if the load resistance $R = 1200\,\Omega$ and the internal resistance $r = 1\,\Omega$ then the terminal voltage is

$$V = E \times \frac{1200}{1200 + 1} = E \times \frac{1200}{1201}$$

which is very nearly the value of the e.m.f.

However if the load resistance is small, say $2\,\Omega$, with an internal resistance of $1\,\Omega$, then

$$V = E \times \frac{2}{2 + 1} = E \times \frac{2}{3}.$$

That is, one third of the potential e.m.f is 'lost' inside the battery.

Self-assessment questions

In the following questions indicate whether the statements are true or false.

40   Internal resistance may be defined as the resistance offered by the source to the flow of current.

TRUE/FALSE

41   When a cell is supplying current, the internal resistance of the cell has no great effect on the output voltage.

TRUE/FALSE

Complete the following statements:

42   The p.d. measured across the terminals of a battery when there is no load connected to it is called the_____ .

43   The p.d. measured across the terminals of a battery when there is a load connected to it is called the_____ .

44   The resistance which a source of electrical energy offers to the current it produces is called the_____.

45   The equation relating E, R and r and the terminal voltage V is given by,
$V =$ _____ .

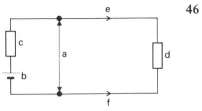

Figure 49   *Circuit for self-assessment question 46*

46 The circuit shown in Figure 49 has parts lettered (*a*) to (*f*). Identify each of these parts using the list of terms numbered 1 to 7. Any of these terms may be used more than once.

(*a*)
(*b*)
(*c*)
(*d*)
(*e*)
(*f*)

(1) current flow
(2) electromotive force
(3) electron flow
(4) internal resistance
(5) load resistance
(6) terminal voltage
(7) cell

47 Sketch the circuit for a cell of e.m.f. 12 V, internal resistance 0.8 Ω, connected in series with a resistor of value 50 Ω.

48 For the circuit described in Question 47, calculate:
(*a*) the terminal voltage V,
(*b*) the circuit current I,
(*c*) the voltage drop across the battery.

49 For the circuit shown in Figure 50 calculate:
(*a*) the total resistance of the load resistors,
(*b*) the terminal voltage V,
(*c*) the circuit current I,
(*d*) the voltage drop V_r across the cell.

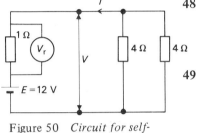

Figure 50   *Circuit for self-assessment question 49*

Solutions to self-assessment questions

40   TRUE.

41   FALSE. The internal resistance causes a voltage drop across the cell.

42   e.m.f.

43   Terminal voltage.

44   Internal resistance.

45 $V = E \times \dfrac{R}{R + r}$.

46   (*a*)   terminal voltage,
(*b*)   cell,
(*c*)   internal resistance,
(*d*)   load resistance,
(*e*)   current flow,
(*f*)   electron flow.

47   See Figure 51.

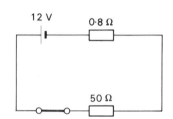

48   (*a*) $V = E \times \dfrac{R}{R + r} = 12 \times \dfrac{50}{50.8} = 11.81$ V.

(*b*) $I = \dfrac{V}{R} = \dfrac{11.81}{50} = 0.236$ Ω.

(*c*)   Voltage drop $= Ir = 0.236 \times 0.8 = 0.19$ V.

49   (*a*)   For parallel resistors $\dfrac{1}{R} = \dfrac{1}{4} + \dfrac{1}{4} = \dfrac{2}{4}$. ∴ $R = 2$ Ω.

(*b*) $V = E \times \dfrac{R}{R + r} = 12 \times \dfrac{2}{(2 + 1)} = 8$ V.

(*c*) $I = \dfrac{V}{R} = \dfrac{8}{2} = 4$ A.

(*d*) $V_r = I \times r = 4 \times 1 = 4$ V.

12 V   0·8 Ω

50 Ω

Figure 51   *Solution to self-assessment question 47*

Section 2

Conductors in the magnetic field

After reading the following material, the reader shall:

2   Know the factors influencing the force on a conductor in a magnetic field and their use in meters and motors.

2.1 Identify the accepted method of representing the direction of flow of conventional electric current in a single conductor.

2.2 Identify the accepted method of representing the direction of the magnetic field produced by a current carrying conductor.

2.3 Sketch the type of magnetic field produced by a solenoid.

Previous studies have included the topic magnetism. Magnetism, like electricity cannot be seen. However, the effects it produces can be seen. A *magnetic field* surrounds a magnet in all directions, the field strength being greatest at the poles of the magnet. When suspended freely and horizontally, a magnet sets itself north-south parallel to the earth's magnetic field. The ends of the magnet are called the poles of the magnet. The *north-seeking pole* is the end of the magnet which points north if the magnet is suspended freely; the *south-seeking pole* is the opposite end.

Representations of the magnetic fields around (i) a bar magnet, and (ii) a horseshoe magnet, are shown in Figure 52 (*a*) and (*b*) respectively. The invisible lines of force leaving the magnet at one point and entering at another are called *flux lines*, and the shape they take up is called the *flux pattern* or *magnetic field*. The amount of flux per unit area is a measure of the *flux density*.

The phenomena associated with magnetic poles are
    (i) *unlike poles attract*,
and (ii) *like poles repel*.

Figures 53 to 56 show the conventions of polarity and direction of flux. It should be noted that *flux lines never cross*. They always distort to prevent this happening; see Figure 57. There is no known insulator for magnetic flux. Magnetic flux passes through any material.

In 1820 a link between electrical energy and magnetic energy was discovered. It was found that a magnetic field occurred around a conductor connected to the terminals of a battery. This magnetic field can be illustrated by setting up the demonstration illustrated in Figure 58. The conductor and perspex sheet are supported by insulated clamps to a stand. The conductor passes through a hole in the perspex sheet. On the perspex sheet and around the conductor are placed small com-

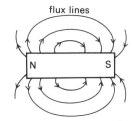

flux lines

magnetic field of bar magnet

(*a*)

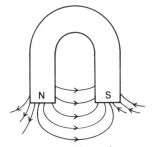

magnetic field of horseshoe magnet

(*b*)

Figure 52   *Magnetic fields round a bar magnet and a horseshoe magnet*

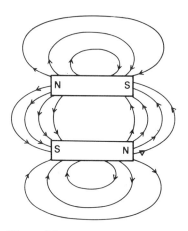

Figure 53   *Magnetic forces tending to pull the magnets together*

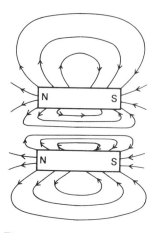

Figure 54   *Magnetic forces tending to push the magnets apart*

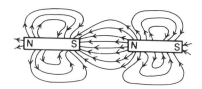

Figure 55   *Unlike poles attract*

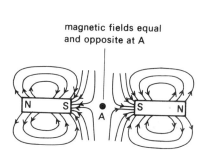

Figure 56   *Like poles repel*

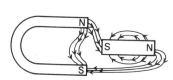

Figure 57   *Paths of magnetic flux*

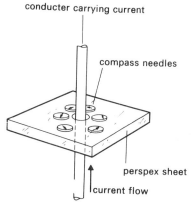

Figure 58   *Demonstration of magnetic field around a conductor*

passes. When the conductor is connected to a source of e.m.f., current flows in the conductor and the compass needles are deflected. A line joining the compass needle directions together forms a circle around the conductor. When the direction of the current flow is reversed the compass needles reverse their direction.

Note that, when describing the directions of current flow in conductors and their associated magnetic fields, it is usual to use the *conventional direction of current flow*, i.e. the current flow is from positive terminal to negative terminal.

conventional current flowing away from the observer

(a)

conventional current flowing towards the observer

(b)

Figure 59 *Conventional current representation*

Figure 59 shows the method of representation for conventional current flow. The conductor is represented by a circle. When the direction of current flows inwards (i.e. away from the observer), a cross is placed in the circle representing the conductor. When the direction of conventional current flow is outwards (i.e. towards the observer), a dot is placed in the circle representing the conductor.

Figure 60 illustrates the representation used for the direction of current flow and magnetic field. In Figure 60 (*a*) and (*b*) the magnetic field is represented by concentric circles. The direction taken by the compass needles is represented by an arrow on the circle representing the magnetic field. This represents the direction of the magnetic field around the conductor. Thus as shown in Figure 60 (*a*), *the direction of the magnetic field is clockwise for an inward flow of conventional current. The direction of the magnetic field is anticlockwise for an outward flow of conventional current*, as shown in Figure 60 (*b*).

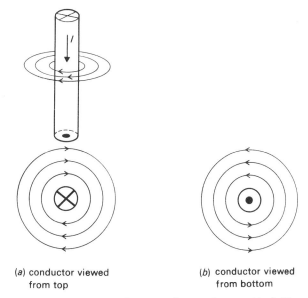

(a) conductor viewed from top

(b) conductor viewed from bottom

Figure 60   *Representation of current flow and magnetic field*

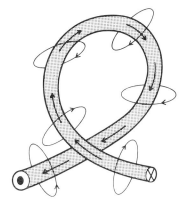

Figure 61   *Magnetic field around a single loop*

A strong magnetic field can be produced by passing a current through a multi-turn coil. Figure 61 illustrates the magnetic field that is produced when a current passes through a single loop. Figure 62 (*b*) illustrates a coil made up of several loops. Figure 62 (*a*) represents this coil sectioned through the top and bottom of each loop, with the direction of the conventional current flow represented by the convention shown in Figure 59. When the current is passed through the coil, a magnetic field is set up around each loop of the coil (Figure 61). Figure 63 shows the magnetic field set up by the current in each conductor, and the field

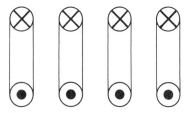

Figure 62 (*a*)   *Current flow through coil*

Figure 62 (*b*)   *Representation of current flow in a coil*

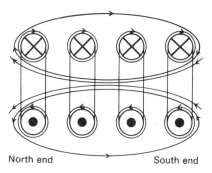

North end   South end

Figure 63   *Magnetic field around coil*

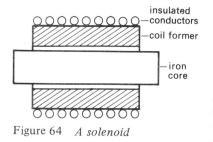

insulated
conductors
coil former

iron
core

Figure 64   *A solenoid*

produced by all the conductors in the coil. This field is represented by magnetic lines of force, which leave one end of the coil, and return to the other end. The end of the coil from which the lines of magnetic force leave has the same magnetic effects as the north-seeking pole of a permanent magnet. The end of the coil where the lines of magnetic force enter behaves as if it were the south-seeking pole of a permanent magnet.

The magnetic flux and the flux density, produced by the current in the coil, can be controlled by altering the current in the coil. An increase in the current increases the flux and the flux density. When the current is switched off the flux very quickly returns to zero.

For a given number of turns in the coil, and for a given current in the turns, the total flux is constant. The flux density can be increased by concentrating the flux into a smaller area. This can be done by introducing an iron core into the centre of the coil, the arrangement being known as a *solenoid* or *electromagnet*. Figure 64 shows a section through a solenoid. The iron core is designed so that when the current in the coil changes or is switched off, there is an almost instantaneous change in flux density in the core.

Engineers in the nineteenth and twentieth centuries have used electromagnets for a wide variety of purposes, e.g. in electric motors, switches, telephones and tape recorders. In each application current flows through a coil with an iron core, and the core acts like a magnet to attract other ferrous metals.

Self-assessment questions

50   Figure 65 shows a length of conductor carrying a current. Sketch the symbol to represent the direction of flow of conventional current when viewed from A and when viewed from B.

51   Figures 66 (*a*) and (*b*) show the magnetic field round a current-carrying conductor. Show the direction of conventional current flow required to produce the direction of magnetic field shown.

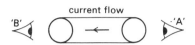

Figure 65   *Current carrying conductor*

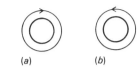

Figure 66   *The direction of the magnetic field in a single conductor*

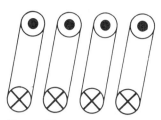

Figure 67   *Sectioned coil*

52   Figure 67 represents the section through a coil with a conventional current passing through it.
For this coil:
(a)   draw the path taken by the lines of magnetic force in and around the whole coil,
(b)   indicate the direction of the lines of magnetic force,
(c)   indicate the magnetic polarity at each end of the coil.

After reading the following material, the reader shall:

2.4   Sketch the resulting magnetic field flux developed by a current-carrying conductor in a permanent magnetic field.
2.5   State that a current-carrying conductor is subjected to a force when it is in a magnetic field.
2.6   Deduce the direction of the force acting on a current-carrying conductor in a magnetic field.
2.7   Use Fleming's left hand rule.
2.8   State that the force on a current-carrying conductor in a magnetic field depends upon:
(a)   intensity of the field,
(b)   strength of the current,
(c)   length of conductor perpendicular to the magnetic field.
2.9   Use the formula $F = B\ell I$.
2.10   Sketch and label the movement of a moving-coil instrument.
2.11   Explain the action of a moving-coil instrument.
2.12   State the major function of:
(a)   the permanent magnet,
(b)   the iron core,
(c)   the coil,
in a moving-coil instrument.
2.13   State that the deflecting force in a moving-coil instrument is directly proportional to the value of the current flowing in the coil.
2.14   State that the moving-coil instrument in an unmodified form is used only with a d.c. supply.

A conductor carrying a current is surrounded by a magnetic field. If the conductor lies in another magnetic field between the poles of a magnet, the two fields must interact. Because the lines of flux never cross, the two fields must either crowd together or cancel each other out, so producing either strong or weak resultant fields. Figures 52 to 56 show this crowding and cancelling effect on the fields.

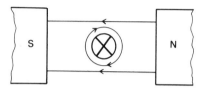

Figure 68   *Interacting fields shown separately*

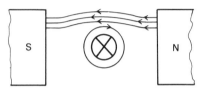

Figure 69   *Lines acting in opposite directions produce a weaker field*

Figure 70   *Lines acting in the same direction produce a stronger field*

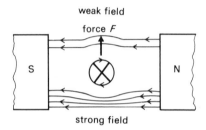

Figure 71   *Resultant interacting field*

Solutions to self-assessment questions

50   View from A represented by the symbol shown in Figure 59 (*a*).
View from B represented by the symbol shown in Figure 59 (*b*).

51   (*a*)   Current is flowing 'in'; therefore the conductor is marked as shown in Figure 59 (*a*).
(*b*)   Current is flowing 'out'; therefore the conductor is marked as shown in Figure 59 (*b*).

52   The diagram should be similar to that shown in Figure 72.

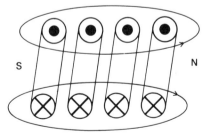

Figure 72   *Solution to self-assessment question 52*

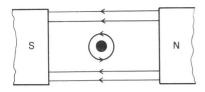

Figure 73 *Separated interacting fields*

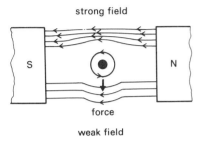

Figure 74 *Combined interacting fields*

Figure 68 shows a straight conductor lying at right angles to a magnetic field, the two separate magnetic fields being shown. However, in the space above the conductor, the two fields oppose each other, as shown in Figure 69, and tend to cancel each other out; thus a weaker field is produced. In the space below the conductor the two fields assist one another, as shown in Figure 70, and thus produce a stronger field. Figure 71 shows the resultant interacting field, with the field weakened above, and strengthened below, the conductor. Because there is a difference in field strength below and above the conductor, there is a force acting on the conductor in the direction shown in Figure 71. The direction of the force on the conductor can be changed by altering the direction of current flow or the direction of the magnetic field or flux.

Consider the conductor and magnetic field shown in Figure 73. The interacting magnetic fields are again shown separately. For this arrangement the stronger field is above the conductor and the weaker field is below the conductor. The resultant interacting field is as shown in Figure 74, and the conductor is subjected to a force in the downwards direction as shown.

Self-assessment questions

53 For the conductor–magnet arrangements shown in Figure 75 (*a*) and (*b*), draw the direction of the magnetic field flux for the magnet and the current-carrying conductor.

54 For the conductor–magnet arrangements shown in Figure 75 (*a*) and (*b*), draw the pattern of the resultant interacting field and indicate the direction of the force on the conductor.

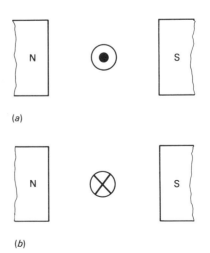

Figure 75 *Conductor-magnet arrangement*

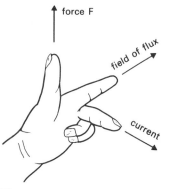

force F

field of flux

current

Figure 76   *Fleming's left-hand rule*

As can be seen from the examples considered, the direction in which the force acts depends upon:

(i) the direction of the magnetic field of flux, and
(ii) the direction of the current flow.

All three directions are at right angles to each other, and any one direction can be determined from a knowledge of the other two by use of *Fleming's left-hand rule*. This is illustrated in Figure 76. The rule states that:

> When the thumb and first two fingers of the left hand are held at right angles to each other in such a way that the first finger is pointed in the direction of the magnetic field of flux, and the second finger is pointed in the direction of the current flow, then the thumb is pointed in the direction of the force *F* acting on the conductor.

Check the direction of the force on the conductor in the previous examples using this rule.

The magnitude of the force *F* on the current-carrying conductor, lying at right angles to a magnetic field of flux, is proportional to three quantities:

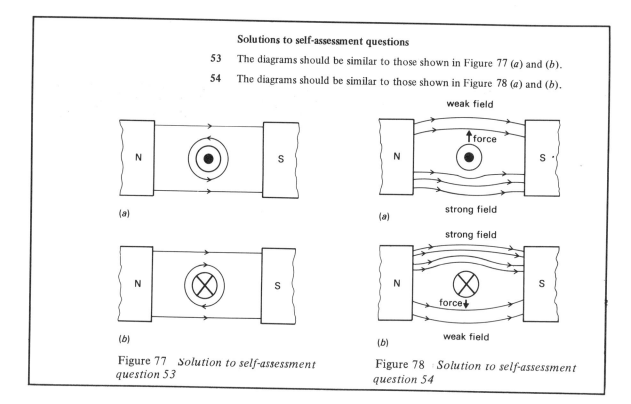

Solutions to self-assessment questions

53 The diagrams should be similar to those shown in Figure 77 (*a*) and (*b*).

54 The diagrams should be similar to those shown in Figure 78 (*a*) and (*b*).

Figure 77   *Solution to self-assessment question 53*

Figure 78   *Solution to self-assessment question 54*

(i) the density of the magnetic flux,
(ii) the magnitude of the current flowing,
(iii) the length of the conductor which is in the magnetic field.

Let B represent the magnetic flux density, the units of which are tesla (T),

ℓ represent the length in metres of the conductor which is lying in the magnetic field,

and I represent the value of the current in amperes flowing through the conductor.

Then $F \propto B\ell I$.

and $F = kB\ell I$ where k is a constant.

By using the units listed above the value of k is unity, and so the equation can be written:

$F = B\ell I$ newtons

Example 11

A conductor carrying a current of 2.5 A is at right angles to a magnetic field, of flux density 0.15 T. The length of the conductor in the magnetic field is 125 mm. Calculate the force on the conductor.

$F = B\ell I$ newtons
$\therefore\ F = 0.15 \times 125 \times 10^{-3} \times 2.5$
$F = 0.047$ N

Example 12

A conductor is carrying a current of 100 A. A 350 mm length of the conductor is in a magnetic field resulting in a force of 40 N being exerted on the conductor. Calculate the flux density of the magnetic field.

From $F = B\ell I$

$$B = \frac{F}{\ell I}$$

$$\therefore B = \frac{40}{350 \times 10^{-3} \times 100} \text{ tesla}$$

$\therefore B = 1.143$ tesla

An important application of the idea of interaction between magnetic fields of flux is in electrical measuring instruments; for example, ammeters and voltmeters. There are two main classes of ammeters and voltmeters in common use. These are the *moving-iron type*, and the *moving-coil type*. Only the moving-coil type of instrument will be considered because this instrument is probably the most widely used. Nearly all modern meters use the moving-coil galvanometer as a basic

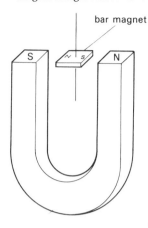

horseshoe magnet

Figure 79   *Suspended magnet*

meter movement, so the principle of the basic meter movement is very important.

The *moving-coil galvanometer* works on the principle of magnetic attraction and repulsion. Figure 79 shows a small bar magnet suspended between the poles of a horseshoe magnet. If the small bar magnet is allowed to turn freely, it takes up the position shown in Figure 79, because *unlike poles attract each other*.

If the bar magnet is turned to any other position (such as that shown in Figure 80) the operator feels the magnet trying to turn back to the position where the unlike poles are exactly opposite, because *like poles repel each other*.

Figure 81 shows the bar magnet still suspended between the poles of the horseshoe magnet, but with the addition of a spring. The spring is tensioned so that at the position where the like poles are exactly opposite, the spring tension is zero. If the bar magnet is allowed to settle in a position of equilibrium, the repulsion forces from the like poles are just balanced by the spring force, and the bar magnet takes up some intermediate position as shown. The greater the magnetic force between the like poles, then the further the bar magnet moves against the spring.

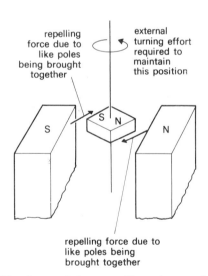

Figure 80   *Suspended magnet like poles together*

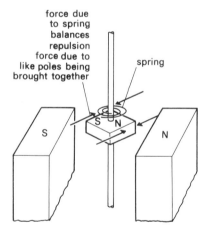

Figure 81   *Suspended magnet with spring*

If the bar magnet is now replaced by a coil of wire as shown in Figure 82, the basic principle of the galvanometer is established. When there is no current flowing through the coil, the spring holds the coil in the position where the like poles are exactly opposite to each other. When an electric current flows through the coil it acts like a magnet, as was

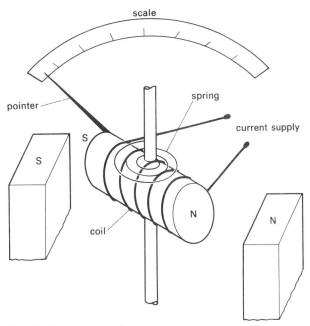

Figure 82   *Basic moving-coil meter*

shown in Figure 63. The magnetic field between the like poles causes the coil to be turned to a position where the spring tension just balances the repulsion force from the like poles. If a pointer is fastened to the coil and allowed to pass over a scale, then the amount of current flowing can be read from the scale. This is the principle on which the basic moving-coil meter works.

In practice, the moving-coil meter is more sophisticated than that shown in Figure 83, but the working principle is the same. The permanent magnet provides the constant magnetic flux. The iron core is fixed between the poles of the permanent magnet. The iron core reduces the air gap between the poles of the magnet, and ensures a uniform flux density in the air gap. The central spindles are mounted on the aluminium former. The spindles are located in high quality bearings in order to reduce friction to a minimum. Hair springs are fastened to each spindle. The hair springs serve several purposes; first to conduct current to and from the coil; secondly to control the deflection of the pointer at the zero position to the scale; finally to restore the pointer to the zero position when the current flow in the coil ceases.

The coil consists of turns of insulated copper conductor wound round the aluminium former. The coil, former and pointer (i.e. all the moving parts) are made as light as possible.

The magnitude of the current in the coil determines the magnitude of the deflection shown by the pointer. The direction of the current flow determines the direction in which the pointer moves because the

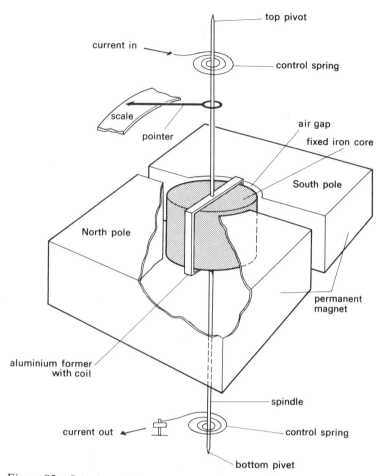

Figure 83   *Principle of the moving-coil instrument*

direction of the permanent magnetic field is fixed. Thus a reversal of direction of current flow causes the coil to move in the opposite direction. For this reason *the moving-coil instrument should be used in d.c. circuits only.*

Self-assessment questions

For each of the following statements the reader should indicate whether he considers them to be true or false:

55   A current-carrying conductor is subjected to a force when it is in a permanent magnetic field.

TRUE/FALSE

56 A moving-coil instrument in its simple form can be used in a.c. and d.c. circuits.

TRUE/FALSE

57 The direction in which the pointer of a moving coil instrument moves is determined by the direction of the current flow.

TRUE/FALSE

Complete each of the following statements:

58 In a moving-coil instrument the major function of the iron core is to ensure_____ .

59 In a moving-coil instrument the purpose of the permanent magnet is to _____ .

60 In a moving-coil instrument the major function of the coil is to _____ .

61 When using Fleming's left hand rule:
the thumb represents the _____ ,
the first finger represents the _____ ,
the second finger represents the_____ .

Figure 84   *Single conductor in magnetic field*

62 Using Fleming's left hand rule determine:
(*a*) the direction of the force *F* on the current-carrying conductor shown in Figure 84.
(*b*) the direction of the current flow in the current-carrying conductor shown in Figure 85.
(*c*) the polarity of the permanent magnet shown in Figure 86.

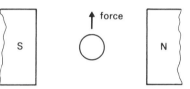

Figure 85   *Force on single conductor in magnetic field*

Figure 86   *Force on single conductor*

63 Calculate the force in newtons acting on a conductor carrying a current of 100 A in a magnetic field of flux density 2 T, if the length of conductor in the magnetic field is 250 mm.

64 The active length of a conductor carrying a current of 50 A at right angles to a magnetic field is 500 mm. If the force on the conductor is 20 N, find the flux density of the magnetic field.

65 A straight conductor carries a current of 200 A at right angles to a uniform magnetic field having a flux density of 0.1 T. Calculate the force exerted on the conductor in newtons per metre.

66   Figure 88 shows a sketch of a moving-coil instrument. Label the parts lettered A to J.

67   State the three factors which affect the value of the force developed on a current-carrying conductor in a magnetic field.

68   Figures 89 (*a*), (*b*), (*c*) and (*d*) show current-carrying conductors in permanent magnetic fields. Sketch the resulting magnetic field flux developed, and hence deduce the direction of the force acting on the current-carrying conductor.

Solutions to self-assessment questions

55   TRUE.

56   FALSE. The direction of movement of the pointer depends upon the current. If the polarity of the supply is reversed the pointer will move in the opposite direction.

57   TRUE.

58   Uniform flux density.

59   Provide the constant magnetic flux.

60   Act as a magnet.

61   Thumb represents the direction of the force acting on the conductor,
First finger represents the direction of magnetic field of flux,
Second finger represents the direction of the current.

62   (*a*)   See Figure 87 (*a*).
(*b*)   See Figure 87 (*b*).
(*c*)   See Figure 87 (*c*).

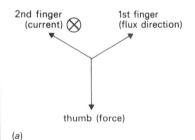

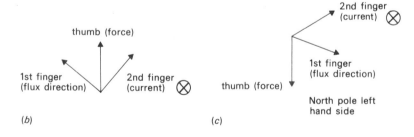

(*a*)
Figure 87   *Solution to self-assessment question 62*

63    $F = B\ell I = 2 \times (250 \times 10^{-3}) \times 100 = 50\ \text{N}$

64    $F = B\ell I$ rearrange to give $B = \dfrac{F}{\ell I}$,

$B = \dfrac{20}{(500 \times 10^{-3}) \times 50} = 0.8\ \text{T}$

65    $F = B\ell I$ where $\ell = 1$ metre
$F = 0.1 \times 1 \times 200 = 20\ \text{N/m}$.

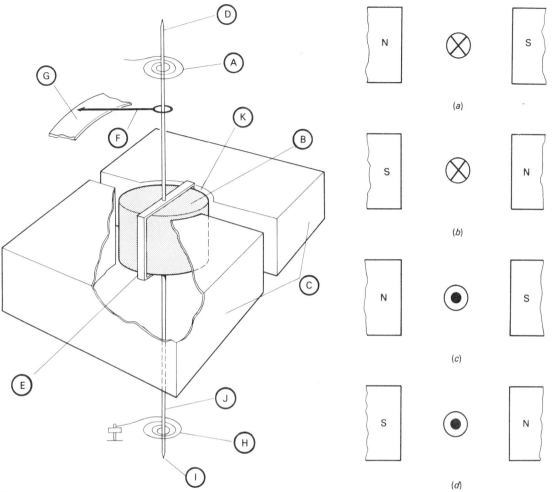

Figure 88   *Self-assessment question 66*

Figure 89   *Self-assessment question 68*

After reading the following material, the reader shall:

2.15   Explain the action of a simplified d.c. motor.

2.16   State that a d.c. motor operates on the same principle as a galvanometer movement.

2.17   Sketch the magnetic field produced by the single-loop armature coil acting in a magnetic field.

2.18   Identify the following basic parts of a d.c. motor:
(i)   d.c. supply,
(ii) pole pieces,
(iii) armature coil,
(iv) commutator,
(v)   brushes.

2.19   State the main function of the basic parts of a d.c. motor.

The basic meter movement of the moving-coil meter has been discussed in some detail. A d.c. motor operates on the same principle as a galvanometer movement.

Solutions to self-assessment questions

66   A – control spring   F – pointer
         B – fixed iron core   G – scale
         C – permanent magnet   H – control spring
         D – top pivot   I – bottom pivot
         E – aluminium former with   J – spindle
              coil

67   (i)   the intensity of the magnetic field,
         (ii) the strength of the current,
         (iii) the length of the conductor perpendicular to the magnetic field.

68   Figure 90 illustrates the resulting magnetic field and shows the direction of the force.

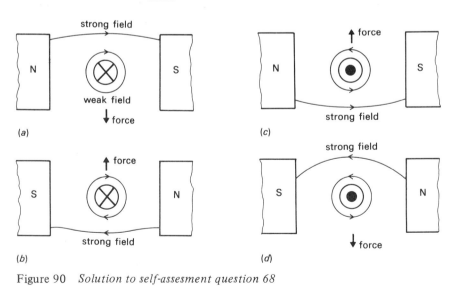

Figure 90   *Solution to self-assesment question 68*

Figure 91 illustrates the elementary d.c. motor. It consists of a loop of wire between the poles of a magnet. The ends of the loop connect to commutator segments, which in turn make contact with the brushes. The brushes have connecting wires leading to a source of d.c. For the purposes of description the two sides of the loop have been lettered A and B as shown.

Figure 92 shows the elementary d.c. motor reduced to a coil rotating in a magnetic field. With the current directions as shown, the magnetic field set up round the coil is equivalent to a magnet with the north-seeking pole at the top position. The magnetic poles of the loop are naturally attracted to the opposite poles of the field magnet. Hence, the loop rotates in a clockwise direction.

When the loop has rotated to a position 90° from the starting point, that is, in the position shown in Figure 93, the current through the loop is reversed. This process is called *commutation*, and occurs when the commutator in contact with the brush connected to the negative terminal of the d.c. supply, suddenly comes into contact with the brush connected to the positive terminal of the d.c. supply. The effect of reversing the current passing through the loop is to reverse the magnetic field generated by the loop. Thus the loop becomes in effect a magnet with the north-seeking pole opposite the north pole piece. Because like poles repel, the effect is to keep the loop rotating in a clockwise direction.

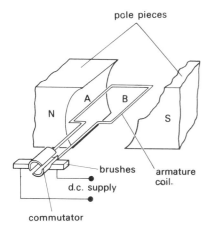

Figure 91　*Elementary d.c. motor*

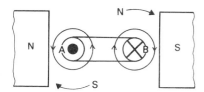

clockwise movement due to unlike poles attracting each other

Figure 92　*Loop in start position*

When the loop has reached the position shown in Figure 94, commutation again takes place, and again the situation of like poles being opposite each other occurs. The repelling effect of the like poles keeps the loop rotating in a clockwise direction.

This is the fundamental action of the d.c. motor. The commutator causes the current through the loop to reverse at the instant unlike poles are facing each other. This causes a reversal in the polarity of the field. Hence, repulsion exists instead of attraction and the loop continues to rotate.

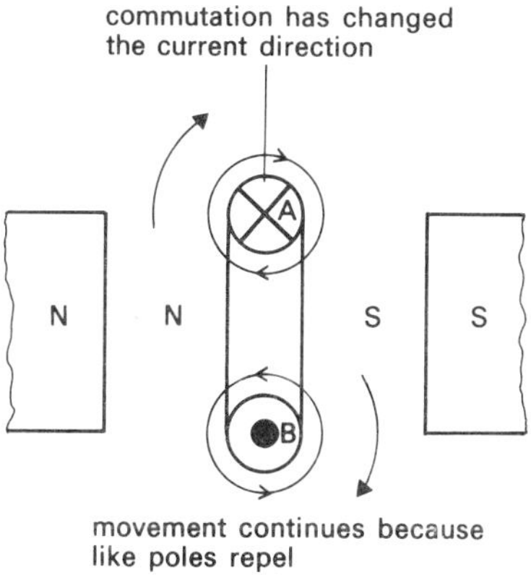

commutation has changed the current direction

movement continues because like poles repel

Figure 93　*Loop position 90°* *from start position*

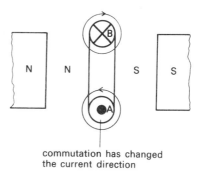

commutation has changed the current direction

Figure 94　*Loop position 270° from start position*

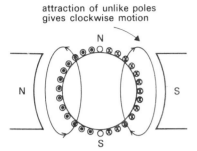

attraction of unlike poles
gives clockwise motion

Figure 95   *Magnetic field in
d.c. motor*

In practice, the armature has many coils instead of just the one loop. This causes the armature winding to act like a coil whose axis is perpendicular to the main magnetic field. Figure 95 illustrates the magnetic field set up by the armature winding. The north-seeking pole of the armature field is attracted to the south-seeking pole of the main field. Thus the armature moves in a clockwise direction. Because of the large number of coils, a continuous turning force is maintained on the armature.

Self-assessment questions

Complete each of the following statements:

69   The d.c. motor works on the same basic principle as the _____
_____.

70   The rotating part of a d.c. motor is called the_____ .

71   In a simple d.c. motor the function of the commutator is to _____
_____ .

72   In a simple d.c. motor the pole pieces provide the _____ .

73   In a simple d.c. motor, every time unlike poles are opposite each other the polarity of the armature current is_____.

74   Figure 96 shows a diagrammatic arrangement of a simple d.c. motor. Name the parts numbered 1 to 5.

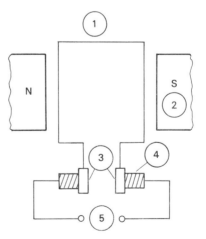

Figure 96   *Simplified d.c. motor*

75 Match the following component names with their main function in a d.c. motor, by placing a 'function' letter in the brackets provided:

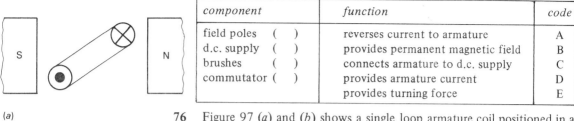

component	function	code
field poles ( )	reverses current to armature	A
d.c. supply ( )	provides permanent magnetic field	B
brushes ( )	connects armature to d.c. supply	C
commutator ( )	provides armature current	D
	provides turning force	E

(a)

(b)

Figure 97 *Single loop armature coil in magnetic field*

76 Figure 97 (*a*) and (*b*) shows a single loop armature coil positioned in a permanent magnetic field. For each case shown:

(i) sketch the magnetic field produced by the current-carrying coil,

(ii) state whether the resulting motion of the coil is clockwise or anti-clockwise, giving the reason for your choice.

Solutions to self-assessment questions

69   The d.c. motor works on the same basic principle as the moving-coil galvanometer.

70   The rotating part of a d.c. motor is called the armature.

71   In a simple d.c. motor the function of the commutator is to reverse the direction of the armature current.

72   In a simple d.c. motor the pole pieces provide the permanent magnetic field.

73   In a simple d.c. motor, every time unlike poles are opposite each other, the direction of the armature current is reversed.

74   (1) armature coil,
      (2) pole pieces,
      (3) commutators,
      (4) brushes,
      (5) d.c. supply.

75   field poles (B)
      d.c. supply (D)
      brushes (C)
      commutator (A)

76   See Figures 98 (*a*) and (*b*).

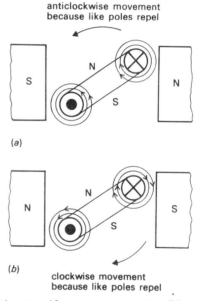

Figure 98   *Solution to self-assessment question 76*

Section 3
Measuring instruments

After reading the following material, the reader shall:

3   Understand the operation, uses and limitations of indicating instruments.

3.1   Describe the use of shunts and multipliers to extend the range of a basic meter movement.

3.2   Calculate the value of the shunt resistance required to extend the working range of a basic meter.

3.3   Calculate the value of the multiplier resistance required to extend the working range of a basic meter.

3.4   Define the sensitivity of an instrument as the ability to detect small current values.

3.5   State that sensitivity is expressed as $\dfrac{1}{\text{f.s.d. current}}$

3.6   State that sensitivity is measured in ohms per volt.

3.7   Use the sensitivity of an instrument in calculations.

3.8   Describe with the aid of diagrams the principles of operation of an ohm-meter.

3.9   Use an ohm-meter for the measurement of resistance.

3.10   Use ammeters and voltmeters in d.c. circuits.

This section of material deals with the use of indicating instruments in direct current circuits. There are three quantities to be measured; these are the *current*, the *potential difference* and the *resistance*. The instruments used to measure these three quantities are the *ammeter*, the *voltmeter* and the *ohm-meter*.

Many modern meters work on the same principle as the moving-coil galvanometer. This principle has already been discussed. The construction of the instrument is illustrated in Figure 83.

Measurement of current
In order to measure the amount of current passing through a component in a d.c. circuit, *an ammeter is connected in series with the component*, as illustrated in Figure 99. The ammeters A_1, A_2 and A_3 measure the current passing through components R_1, R_2 and R_3 respectively.

In the moving coil instrument the current to be measured has to pass through the moving coil, and for the instrument to be sensitive, the movement must be as light as possible. Therefore the number of turns and the size of the wire on the coil must be limited. A typical moving-coil instrument gives a full scale deflection of 15 mA when connected to a 75 mV supply. To enable higher currents to be measured, a moving-coil meter must be used in conjunction with a *shunt*. A shunt is

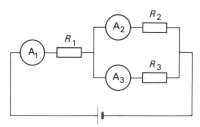

Figure 99   *Ammeter position for current measurement*

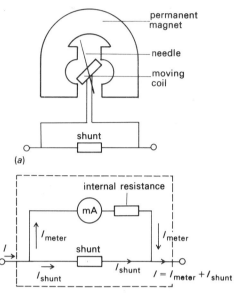

(a)

(b)

Figure 100   *Ammeter shunt arrangement*

a low resistance which is connected in parallel with the meter. The resistance of the shunt is designed so that the maximum current which can flow through the instrument is the current required to produce full scale deflection (f.s.d.). Figure 100 (*a*) shows diagrammatically the basic meter connected in parallel with a shunt; Figure 100 (*b*) shows the circuit for the ammeter in more conventional form. How is the value of the shunt resistor calculated? The method used is a practical example of the application of Ohm's law in parallel circuits.

Example 13

Find the resistance of the shunt required to enable an instrument to read 50 A full scale deflection, if the instrument has a resistance of 5 Ω and a full scale deflection of 15 mA.

Figure 101 shows the circuit required.

Let suffix *m* refer to the meter and suffix *SH* refer to the shunt.

$$I_m = 15 \text{ mA}; I = 50 \text{ A and } R_m = 5 \text{ } \Omega.$$

The voltage required to give full scale deflection is V_m,

where $V_m = I_m R_m$

$$= \frac{15}{1000} \times 5,$$

$$= 0.075 \text{ V}$$

The value of the shunt resistance R_{SH} can be calculated from,

$$R_{SH} = \frac{V_m}{I_{SH}}$$

From Figure 101 it is seen that,

$$I_{SH} = I - I_m$$
$$= 50 - 0.015$$
$$= 49.985 \text{ A}$$

Therefore $R_{SH} = \frac{0.075}{49.985}$

$$= 0.001\,500\,4 \text{ } \Omega$$
$$= 1.5004 \text{ m}\Omega.$$

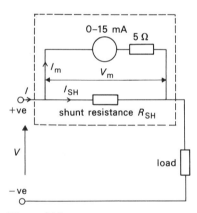

Figure 101   *Ammeter shunt arrangement for Example 13*

Measurement of potential difference

In order to measure the potential difference between two points in a circuit, a voltmeter is used. In Figure 102 the voltmeter V_1 measures the potential difference between points A and B, and the voltmeter V_2 measures the potential difference between points C and D. Note

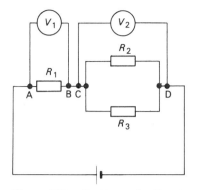

Figure 102 *Position of voltmeters for measuring p.d.*

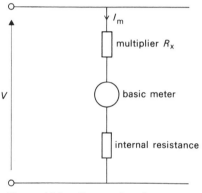

Figure 103 *Connection for multiplier and basic meter*

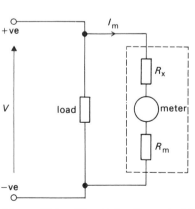

Figure 104 *Circuit for Example 14*

that the voltmeters are connected in parallel with the relevant components.

If a moving-coil instrument is to be used to measure potential difference, i.e. as a voltmeter, then the current passing through it must be limited to the full scale deflection current, otherwise the meter movement will be damaged.

The current through the instrument is limited to a safe value by connecting a comparatively high resistance in series with it; such a resistor is known as a *multiplier*. Figure 103 shows the connection diagram for a multiplier and a basic meter. The calculation of the value of a multiplier resistance is again a simple application of Ohm's law, this time to a series circuit.

Example 14

Calculate the value of the multiplier resistance required to enable a moving-coil instrument, having an internal resistance of 5 Ω and a full scale deflection of 75 mV, to measure a p.d. of 300 V.

Referring to Figure 104. Let suffix m denote the meter.

$R_m = 5 \ \Omega$, $V_m = 0.075$ V and $V = 300$ V

The maximum current that can be allowed to pass through the meter without causing any damage is I_m.

Using Ohm's law $I_m = \dfrac{V_m}{R_m}$

$$= \frac{0.075}{5}$$

$$I_m = 0.015 \text{ A}$$

To keep the meter current down to this value, the resistance of the voltmeter must be increased by the use of the multiplier resistor R_x. That is, the total resistance of the voltmeter circuit $R = R_x + R_m$.

Now using Ohm's law, $R = \dfrac{V}{I_m}$ where $V = 300$ V

$$\therefore \quad R = \frac{300}{0.015}$$

$$R = 20\,000 \ \Omega$$

Therefore the resistance of the multiplier $R_x = R - R_m$

$$= 20\,000 - 5$$

$$= 19\,995 \ \Omega$$

Sensitivity of an instrument

The sensitivity of an instrument is a measure of its ability to detect small currents. It is expressed as the reciprocal of the maximum current it can detect, i.e. the full scale deflection current.

Thus, sensitivity $= \dfrac{1}{\text{f.s.d. current}}$

Since $I = \dfrac{V}{R}$

then, sensitivity $= \dfrac{1}{I} = \dfrac{R}{V}$

That is, the sensitivity of an instrument is expressed in ohms per volt. Thus an instrument which gives a full scale deflection with a current of 15 mA has a sensitivity of $1/0.015 = 66.67$ Ω/V. If this instrument is to be used to measure 300 V, then its total resistance is $66.67 \times 300 = 20\,000$ Ω (see Example 14). Moving-coil instruments can be made to detect currents of a few microamperes, so it is possible for them to have sensitivities as high as 100 kΩ/V.

Example 15

If a voltmeter has a range of 0–50 V and a sensitivity of 400 Ω/V what is:

(a)   the total resistance of the meter,
(b)   the full scale deflection current for the meter,
(c)   the current taken by the meter when it is reading 20 V?

(a)   Total resistance R = sensitivity $\times$ maximum voltage,

$= 400 \left[\dfrac{\Omega}{V}\right] \times 50\ [V]$

$= 400 \times 50$ Ω

$= 20\,000$ Ω

(b)   Since sensitivity $= \dfrac{1}{\text{f.s.d. current}}$

then maximum current $= \dfrac{1}{\text{sensitivity}}$

$= \dfrac{1}{400}$ amperes

$= 0.0025$ A

$= 2.5$ mA

(c)   For a reading of 20 V, the current taken, by proportion, is $\dfrac{20}{50}$ of full scale reading

$$= \frac{20}{50} \times 2.5 \text{ mA}$$

$$= 1 \text{ mA}$$

Alternatively, current $= \dfrac{\text{voltage}}{\text{resistance}}$

$$= \frac{20}{20\,000}$$

$$= 1 \text{ mA}$$

Example 16

A moving-coil instrument has a full scale deflection of 30 mV and an internal resistance of 6 Ω. Calculate the value of the multiplier resistor required to enable the meter to read 600 V.

Using Ohm's law, the full scale deflection $(I) = \dfrac{V}{R}$

i.e. $I = \dfrac{30}{1000} \times \dfrac{1}{6}$

$\quad I = 5 \text{ mA} = 0.005 \text{ A}$

The meter sensitivity $= \dfrac{1}{0.005}$

$$= 200 \ \Omega/\text{V}$$

The total resistance of the meter $=$ sensitivity $\times$ maximum voltage,

$$= 200 \left[\frac{\Omega}{V}\right] \times 600 \ [V]$$

$$= 120\,000 \ \Omega$$

Since total resistance of meter = multiplier resistance + meter resistance, then multiplier resistance $= 120\,000 - 6$

$$= 119\,994 \ \Omega$$

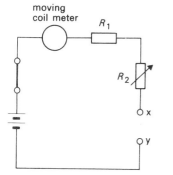

Figure 105   *The simple ohm-meter*

Measurement of resistance

The ohm-meter, as the name suggests, is an instrument for measuring resistance. Figure 105 shows a simple ohm-meter circuit which consists of a fixed resistor R_1, a variable resistor R_2, a moving-coil instrument and a battery, all connected in series. To set the meter, the terminals X and Y are brought together, i.e. they are 'shorted out'; the resistor R_1 protects the meter. The variable resistor R_2 is then adjusted to give full scale deflection on the meter, which corresponds to zero resistance between the terminals. If a resistor is now connected between the terminals X and Y, the current flowing in the circuit must be less than the full scale deflection current, so giving less deflection on the meter.

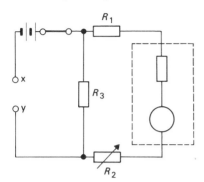

Figure 106   *Ohm-meter circuit –*
extension of range

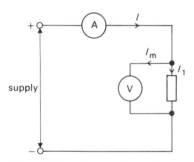

Figure 107   *Connections of*
ammeter and voltmeter

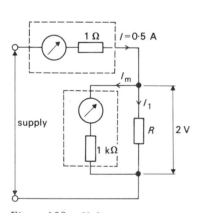

Figure 108   *Voltmeter-ammeter*
method of resistance measurement

The scale of the meter can be calibrated in ohms by connecting standard resistors across the terminals X and Y, and marking the scale accordingly. The scale is non-linear, being cramped at the high resistance end of the scale. This type of ohm-meter does not have a high degree of accuracy; every time the instrument is used, its terminals must be shorted and the zero reading checked. The zero setting should also be checked periodically during prolonged use of the instrument. In practice, all the components shown in the circuit in Figure 105 are contained in one portable case.

The range of the ohm-meter may be changed by using the circuit shown in Figure 106. Resistor R_3 is connected in parallel with the basic ohm-meter circuit. The working range of the ohm-meter can be varied by varying the value of resistor R_3.

To measure the resistance of a component using an ohm-meter, the component is placed between the X and Y terminals of the ohm-meter. The correct scale is selected, if an approximate value is known, and the ohm-meter switched on. The value of the resistance may then be read off the scale. If an approximate value is not known then the highest scale should be selected first and reduced until a suitable scale reading is obtained. If the component is connected into a circuit, *it must be isolated* before its resistance is measured.

When an ammeter or voltmeter is connected into a citcuit to measure the current or p.d., then the reading obtained may not be exact for a number of reasons. Some of these sources of error will be examined in detail in the next section. At this stage it is necessary to consider the possible errors in instrument readings that can result from the use of the instruments themselves. Consider the circuit shown in Figure 107. If the voltmeter reading is V volts and the ammeter reading is I amperes, an approximate value of the resistance R can be obtained from the equation

$$R = \frac{V}{I}$$

This answer can only be approximate because the ammeter reading also includes the current taken by the voltmeter, i.e.

$$I = I_1 + I_m$$

Example 17
A voltmeter having a resistance of 1 kΩ and an ammeter having a resistance of 1 Ω are used to measure an unknown resistance R, as shown in Figure 108. Find the value of the resistor R, and compare it with the value calculated from the voltmeter reading of 2 V and the corresponding ammeter reading of 0.5 A.

The current taken by the voltmeter, using Ohm's law is

$$I_m = \frac{V}{R} = \frac{2}{1000}$$

$$I_m = 0.002 \text{ A}$$

The current passing through the resistor is

$$
\begin{aligned}
I_1 &= I - I_m \\
&= 0.5 - 0.002 \\
&= 0.498 \text{ A}
\end{aligned}
$$

Using Ohm's law, the value of the resistor is

$$R = \frac{V}{I_1}$$

$$R = \frac{2}{0.498}$$

$$R = 4.016 \ \Omega$$

Taking the meter readings and using Ohm's law, the calculated value of the resistance of the resistor is

$$R_c = \frac{2}{0.5}$$

$$R_c = 4 \ \Omega$$

The percentage error $= \dfrac{0.016}{4.016} \times 100$

$$= 0.4\%$$

Note that the supply p.d. is the p.d. across the resistor plus the p.d. across the ammeter. That is,

$$
\begin{aligned}
\text{supply voltage} &= V + IR \\
&= 2 + 0.5 \times 1 \\
&= 2.5 \text{ V}
\end{aligned}
$$

The two values for the resistance compare very favourably. However, consider the same components connected differently, as shown in Figure 109. The voltmeter reads the full supply p.d. of 2.5 V. The total resistance of the branch containing the ammeter and the resistor R is $4.016 + 1 = 5.016 \ \Omega$.

The ammeter reading, using Ohm's law is

$$I = \frac{V}{R}$$

$$I = \frac{2.5}{5.016}$$

$$I = 0.498 \text{ A}$$

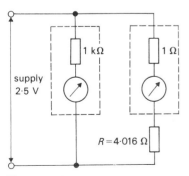

supply
2·5 V

1 kΩ   1 Ω

$R = 4 \cdot 016 \ \Omega$

Figure 109   *Alternative voltmeter-ammeter method of resistance measurement*

Taking the meter readings, the calculated value for the resistor is

$$R = \frac{V}{I}$$

$$= \frac{2.5}{0.498}$$

$$= 5.02 \ \Omega$$

$$\text{percentage error} = \frac{5.02 - 4.016}{4.016} \times 100$$

$$= \frac{1.004}{4.016} \times 100$$

$$= 25\%$$

This time, the two values do not compare very well. If it is required to measure the value of an unknown resistance using the voltmeter-ammeter method, then the circuit shown in Figure 108 should be used.

Example 18

A voltmeter and a resistor are connected as shown in Figure 110 (*a*). Calculate the current which flows:

(*a*)   in the voltmeter,
(*b*)   in the resistor R,
(*c*)   in the circuit.

If an ammeter is now connected as shown in Figure 110 (*b*), calculate the current which flows in the resistor. Express as a percentage the difference between the original current and the current when an ammeter is connected.

(*a*)   Current flowing through voltmeter $= \dfrac{V}{R_1} = \dfrac{12}{400} = 0.03 \ A$

(*b*)   Current flowing through resistor $= \dfrac{V}{R_2} = \dfrac{12}{1} = 12 \ A$

(*c*)   Circuit current $= I_1 + I_2 = 12 + 0.03 = 12.03 \ A$

When the resistor and the ammeter are in series,
total resistance $= R_2 + R_A$
$$= 1 + 0.1$$
$$= 1.1 \ \Omega$$

Therefore the current flowing through the resistor

$$= \frac{12}{1.1} = 10.91 \ A$$

Difference in current values $= 12.03 - 10.91 = 1.12 \ A$

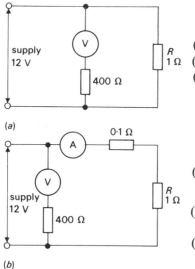

(a)

(b)

Figure 110   *Example 18*

Therefore percentage difference $= \dfrac{1.12}{12.03} \times 100$

$= 9.3\%$

Self-assessment questions

Complete the following statements:

77 A shunt is used to extend the range of _____ .

78 A multiplier is used to extend the range of _____ .

79 The sensitivity of an instrument is a measure of its ability to _____ _____ .

80 The sensitivity of an instrument is expressed as the reciprocal of the _____ .

81 The sensitivity of an instrument is measured in units of _____ .

82 The leads of an ohm-meter are short circuited together when the ohm-meter is being _____ .

83 When measuring the resistance of any device in a d.c. circuit using an ohm-meter, one terminal of the device must be _____ .

84 Sketch the circuit diagrams of a basic meter (*a*) with a shunt, and (*b*) with a multiplier.

85 An instrument has an f.s.d. of 5 mA, and an internal resistance of 5 Ω. Find the resistance of the shunt required to extend the range of the instrument to 50 A f.s.d.

86 A moving-coil instrument has an internal resistance of 5 Ω and a full scale deflection of 75 mA. Calculate the value of the multiplier resistance required to enable the instrument to read up to 150 V.

87 A moving-coil instrument has full scale deflection values of 25 mV and 5 mA, and an internal resistance of 5 Ω. Calculate the sensitivity of the instrument, and the value of the resistances required to enable the instrument to read (*a*) 300 V, and (*b*) 75 A.

88 If a voltmeter has a range of 0–100 V and sensitivity of 500 Ω/V, calculate:
(*a*) the total resistance of the meter,
(*b*) the full scale deflection current for the meter,
(*c*) the current taken by the meter when it is reading 60 V.

89   Figure 111 shows a basic ohm-meter circuit. The various elements are numbered 1 to 7. Match the numbers of the various elements with the functions listed opposite.

Solutions to self-assessment questions

77   An ammeter.

78   A voltmeter.

79   Detect small current values.

80   Full scale deflection current.

81   Ohms per volt.

82   Zeroed.

83   Open circuited.

84   The circuit diagrams should be similar to those shown in Figures 100 and 103. The shunt is connected in parallel with the meter and the multiplier is connected in series with the meter.

85   Full scale voltage = $I_m R_m$ = 0.005 × 5 = 0.025 mV
Shunt resistance = V_m / I_{SH} where $I_{SH} = I - I_m$ =   50 − 0.005
                                               =   49.995 A
Therefore shunt resistance = $\dfrac{0.025}{49.995}$ = 0.50005 mΩ.

86   Maximum current = 75 mA
Total resistance R = $R_x + R_m$ = $\dfrac{V}{I_m}$ = $\dfrac{150}{0.075}$ = 2 000 Ω
Therefore multiplier resistance = 2 000 − 5 = 1 995 Ω.

87   Sensitivity = $\dfrac{1}{\text{f.s.d. current}}$ = $\dfrac{1}{0.005}$ = 200 ohms per volt

(*a*)   Total resistance at 300 V =   200 $\left[\dfrac{\Omega}{V}\right]$ × 300 [V]

                                    =   60 000 Ω = $R_x + R_m$
Therefore the multiplier resistance =   (60 000 − 5) Ω
                                    =   59 995 Ω

(*b*)   Current to be carried by shunt = 75 − 0.005 = 74.995 A.
Therefore shunt resistance = $\dfrac{V_m}{I_{SH}}$ = $\dfrac{0.025}{74.995}$ = 0.333 mΩ.

88   (*a*)   Total resistance =   sensitivity × voltage

                                    =   500 $\left[\dfrac{\Omega}{V}\right]$ × 100 [V] = 50 000 Ω

(*b*)   Sensitivity = $\dfrac{1}{\text{f.s.d. current}}$

Therefore f.s.d. current = $\dfrac{1}{\text{sensitivity}}$ = $\dfrac{1}{500}$ = 2 mA

(*c*)   Current = $\dfrac{60}{100}$ × f.s.d. current = $\dfrac{60}{100}$ × 2 mA = 1.2 mA

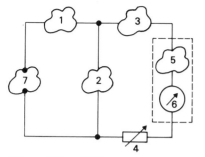

Figure 111   *Self-assessment question 89*

element number	element function
	meter movement and scale
	protects meter when the leads are short circuited
	internal resistance of meter movement
	used to zero ohm-meter before using
	determines the range of the ohm-meter
	internal/independent supply to ohm-meter
	resistance being measured

90 The circuit shown in Figure 112 shows component R_1 connected in series to a supply. The switch can be in the closed or open position. The meters M_1 and M_2 can represent ammeters, voltmeters or ohm-meters. Study the circuit and then complete the following statements.

(i) Before connecting any meter into the circuit, the switch should be in the (OPEN/CLOSED) position.

(ii) The correct meter connection required to measure the resistance of component R_1 is shown by meter _____.

(iii) To measure the resistance of component R_1 the switch should be in the (OPEN/CLOSED) position.

(iv) The correct meter connection required to measure the p.d. across component R_1 is shown by meter _____ .

(v) To measure the p.d. across component R_1, the switch should be in the (OPEN/CLOSED) position.

(vi) The correct meter connection required to measure the current flowing through component R_1 is shown by meter _____.

(vii) To measure the current flowing through the component R_1, the switch should be in the (OPEN/CLOSED) position.

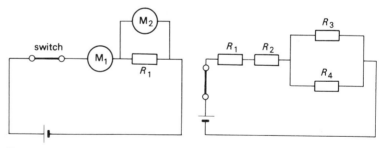

Figure 112   *Self-assessment question 90*

Figure 113   *Self-assessment question 91*

91 On the circuit in Figure 113, indicate the meter positions and connections required to measure:

(*a*) the terminal voltage,

(*b*) the load current,

(*c*) the potential difference across resistor R_2,

(*d*) the current flowing through resistor R_3.

92   Study the circuit in Figure 115 and then complete the statements shown below by inserting one of the following four answers (assume the p.d. of the supply is maintained constant).

(i)   increase,
(ii)   decrease,
(iii) remain the same,
(iv)   become zero.

(i)   If resistor R_1 goes on open circuit, then the reading on the ammeter will _____ .

(ii)   If the resistor R_1 goes on open circuit, then the reading on the voltmeter will _____ .

(iii) If the resistors R_1 and R_2 go on open circuit, then the reading on the ammeter will _____ .

(iv)   If the resistors R_1 and R_2 go on open circuit, then the reading on the voltmeter will _____ .

Solutions to self-assessment questions

89

element number	element function
6	meter movement and scale
3	resistance to protect meter
5	internal resistance of meter
4	variable resistor used to zero ohm-meter
2	resistor which enables range of ohm-meter to be changed
1	internal/independent supply
7	resistance being measured

90   (i)   OPEN. A safety precaution.
       (ii) M_2. The ohm-meter is connected across the terminals.
       (iii) OPEN. The ohm-meter has its own supply; to measure a resistance the component must be isolated from the supply.
       (iv) M_2. The voltmeter is connected across the terminals.
       (v)   CLOSED. A current must be flowing.
       (vi) M_1. The ammeter is connected in series with the component.
       (vii) CLOSED. A current must be flowing.

91   The meter positions are as shown in Figure 114.

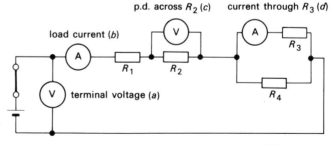

Figure 114   *Solution to self-assessment question 91*

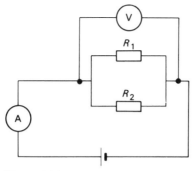

Figure 115   *Self-assessment question 92*

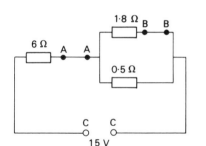

Figure 116   *Self-assessment question 93*

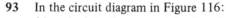

Figure 117   *Self-assessment question 94*

93   In the circuit diagram in Figure 116:
(*a*) calculate the current in the circuit when no instruments are connected,
(*b*) calculate the current when the instruments are connected,
(*c*) express the difference between the two currents as a percentage,
(*d*) calculate the p.d. across the resistor when the instruments are connected,
(*e*) express the difference between the original p.d. and the p.d. when instruments are connected in the circuit as a percentage.

94   In the circuit in Figure 117 when no instruments are connected and the p.d. between C–C is 15 V, calculate:
(*a*) the current in the circuit,
(*b*) the current in the 1.8 Ω resistor.

When an ammeter with a resistor 0.1 Ω is connected between A–A, and the p.d. between C–C is maintained at 15 V, calculate:
(*c*) the current in the circuit,
(*d*) the difference between the original current and the new current as a percentage.

When the ammeter is reconnected between B–B, and the p.d. between C–C is maintained at 15 V, calculate:
(*e*) the current in the 1.8 Ω resistor,
(*f*) express the difference between the original current in this resistor and the new current as a percentage.

After reading the following material, the reader shall:

3.11   Give examples of manufacturing and operating errors.
3.12   Explain the accuracy class index system.
3.13   Define the term fiducial value.
3.14   Calculate the permissible error using the class index and fiducial value.
3.15   Calculate percentage error using the permissible error for different scale values.
3.16   Explain why the accuracy of an instrument varies for different points on the scale.

When using instruments, consideration should be given to the possible sources of error which may result in the incorrect readings being obtained. Errors in the values of readings obtained from measuring instruments can come from a number of sources.

Solutions to self-assessment questions

92   (i)   Decrease. The total resistance is increased; hence there is less current.
     (ii) Remain the same. There is no change in the supply voltage.
     (iii) Become zero. The total resistance is now infinity; there is no current.
     (iv) Remain the same. There is no change in the supply voltage.

93   (a) When there are no instruments in the circuit and the p.d. across the 1 Ω resistor is 6 V, the current in the circuit = $V/R = 6/1 = 6$ A.

     When the ammeter and voltmeter are connected into the circuit, the resistance (R_T) of the circuit is the sum of the resistance of the ammeter and the equivalent resistance (R_E) of the voltmeter resistance and the 1 Ω resistance.

$$\frac{1}{R_E} = \frac{1}{400} + \frac{1}{1} \quad \therefore R_E = \frac{400}{401} = 0.9975 \ \Omega$$

$$\therefore R_T = 1 + 0.9975 = 1.9975 \ \Omega$$

     (b) The new current in the circuit $= \dfrac{V}{R_T} = \dfrac{6}{1.0975} = 5.4669$ A

     (c) The difference between the original current is
          $6 - 5.4669 = 0.5331$ A
          Expressed as a percentage, difference in current $= \dfrac{0.5331}{6} \times 100$

          $= 8.885\%$

     (d) The current flowing in the circuit produces a p.d. across the ammeter. When the resistance of the ammeter is R_I, this p.d. $= I \times R_I = 5.4669 \times 0.1$
          $ = 0.546\,69$ V

          The p.d. across the resistor $= 6 - 0.546\,69$
          $ = 5.453\,31$ V.

     (e) The difference between the circuit p.d. and the p.d. across the resistor, expressed as a percentage is

$$\frac{0.546\,69}{6} \times 100 = 9.1\%$$

94   When there are no instruments connected to the circuit the resistance of the resistors connected in parallel is found from

$$\frac{1}{R_p} = \frac{1}{1.8} + \frac{1}{0.5} \quad \therefore R_p = \frac{0.9}{2.3} = 0.391 \ \Omega$$

     The total resistance of the circuit, $R_T = 6 + 0.391$
     $ = 6.391 \ \Omega.$

     (a) The current in the circuit when the p.d. between C–C is 15 V is

$$\frac{V}{R} = \frac{15}{6.391} = 2.347 \text{ A}$$

(a) *Manufacturing errors*

Because the individual components of any instrument must have been manufactured to certain design tolerances, there is the possibility of some error in the resultant readings from the instrument. These errors

---

The p.d. across the parallel resistors
$$= I \times R_p = 2.347 \times 0.391 = 0.917 \text{ V}$$

(b) The current in the 1.8 Ω resistor $= \dfrac{0.917}{1.8} = 0.5094$ A

When the ammeter is connected in position A–A, the resistance R_n of the circuit is $R_T + R_I$ where R_I is the resistance of the instrument.
$R_n = 6.391 + 0.1 = 6.491 \ \Omega$.

(c) The current in the circuit is $\dfrac{V}{R_n} = \dfrac{15}{6.491} = 2.31$ A.

The difference between the two currents is
$2.347 - 2.31 = 0.037$ A

The introduction of the ammeter into the circuit between A–A has reduced the current in the circuit.

(d) The percentage difference between the two currents

$$= \frac{0.037}{2.347} \times 100 = 1.57\%$$

When the ammeter is connected between B–B, the resistance R_b of the branch which includes the ammeter
$$= 1.8 + 0.1 = 1.9 \ \Omega$$

The resistance of the parallel circuit is found from

$$\frac{1}{R_p} = \frac{1}{1.9} + \frac{1}{0.5} \ . \ \therefore R_p = \frac{0.95}{2.4} = 0.396 \ \Omega$$

The total reistance of the circuit is

$$R_T = 6 + 0.396 = 6.396 \ \Omega$$

(e) The current in the circuit when the p.d. between C–C is 15 V is

$$\frac{V}{R_T} = \frac{15}{6.396} = 2.345 \text{ A}$$

The p.d. across the parallel resistors $= I \times R_p$
$$= 2.345 \times 0.396$$
$$= 0.929 \text{ V}$$

The current in the branch which includes the ammeter is

$$\frac{V}{R_b} = \frac{0.929}{1.9} = 0.488 \text{ A}$$

(f) The difference between the original current flowing in this branch and the new current is

$$0.5094 - 0.488 = 0.0214 \text{ A}$$

The percentage difference $= \dfrac{0.0214}{0.5094} \times 100 = 4.2\%$

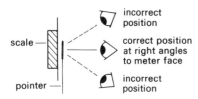

(a) side view

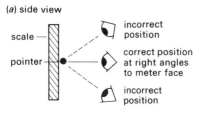

(b) plan view

Figure 118   *Meter viewing position*

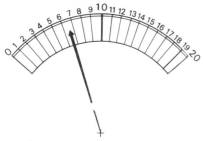

Figure 119   *Estimated reading*

may result from the backlash of gears, friction in bearings or incorrect graduation of scales. However, in the majority of cases, if the instrument is calibrated, i.e. compared against a known standard, the effects of these errors can be removed by adjustment.

(b) Operating errors

One of the most common sources of operating error in instruments which contain a scale and a pointer is the failure of the operator to read the indicated value correctly. Meters should always be read from a position directly in front of the meter face, as shown in Figure 118. The error caused by reading a meter from the wrong angle is called parallax error. The individual meter divisions on the scales are usually small, and the pointer is raised away from the scale in order for it to move freely round it. Thus if the meter is read from a position other than one at right angles to the meter, the resultant reading will be inaccurate, perhaps by as much as one complete division.

Another possible source of error is when a meter reading has to be obtained by estimation. When the pointer of a meter reads a value somewhere in between two divisions on the scale, the amount by which the pointer has moved past the lower division mark on the scale has to be estimated by eye. The amount that is estimated is then added to the value of the lower division mark, to give the reading. Figure 119 shows a pointer on the scale. The value lies between 6 and 7. How far past the 6 division has the pointer moved? An estimated amount would be 0.4 of a division, so the reading is 6.4 units. The value of 0.4 has been estimated by eye.

(c) Environmental conditions

Extremes of temperature or humidity, excessive vibration, corrosive environments and an environment containing a magnetic field are all conditions which may affect the working of an instrument and help to produce a source of error. Such conditions should therefore be avoided whenever possible.

Accuracy

The limits of accuracy of electrical indicating instruments are laid down by the British Standards Institute. The standard for instruments, B S 89:1977, lists the accuracy class of instruments, and conforms with the recommendations of the International Electrotechnical Commission. The accuracy class lists eight class index numbers as indicated in the table opposite.

class index number	0.05	0.1	0.2	0.5	1	1.5	2.5	5
percentage accuracy	±0.05%	±0.1%	±0.2%	±0.5%	±1%	±1.5%	±2.5%	±5%

The class index number indicates the percentage accuracy of the instrument in that class. If an instrument has a class index of 0.05 it has an accuracy of ± 0.05%. However, this accuracy does not apply to all points on the scale, it only applies to the *fiducial value* of the instrument. The fiducial value is the value to which the errors of the instrument are referred. In many cases, but not in every case, the fiducial value is the maximum scale value of the instrument. If the fiducial value is the maximum scale value, then a 0–10 A ammeter would have a fiducial value of 10 A, and a 0–300 V voltmeter would have a fiducial value of 300 V. Reference should be made to B.S.89:1977 for the complete definitions of fiducial value.

The permissible error may be calculated using the equation

$$\text{Permissible error} = \frac{\text{class index}}{100} \times \text{fiducial value}$$

For example, if an ammeter has a class index of, say 0.1, and the fiducial value is the maximum scale value of 10 A, then the permissible error is given by:

$$\text{permissible error} = \frac{0.1}{100} \times 10 = \pm 0.01 \text{ A}$$

That is, at any point on the scale there may be an error of ± 0.01 A, and the instrument will still be in the class index of 0.1. However, this means that the instrument will be less accurate at the lower end of the scale range, because an error of ± 0.01 A represents a greater percentage error on a scale reading of 1 A than it does on a scale reading of 10 A. If the scale reading is 1 A with a permissible error of ± 0.01 A, then the percentage error is

$$\frac{\pm 0.01}{1} \times 100 = \pm 1\%$$

If the scale reading is 10 A with a permissible error of ± 0.01 A, then the percentage error is:

$$\frac{\pm 0.01}{10} \times 100 = \pm 0.1\%$$

Because the accuracy of an instrument varies over the scale range, instrument ranges should be chosen so that the deflection produced is at least 80% of the full scale value which can be measured by the instrument. For example, an ammeter which has a full scale deflection value of 10 A, should not be used for measuring currents less than 8 A.

Instruments with a class index of 0.05 and 0.1 are laboratory reference standards and are used for calibrating other instruments. Class indexes

0.2 and 0.5 include portable instruments, but are still mainly laboratory instruments. Class indexes 1 to 5 include portable instruments, multimeters, switchboard instruments and general purpose instruments. Prior to 1970, instruments were manufactured having only two classes of accuracy, namely Precision (Pr) and Industrial (In). The accuracy of the permanent magnet moving coil instruments to ± 0.3% in the precision range and ± 1% in the industrial range. Reference should be made to BS 89:1954 for further details of these two ranges of instruments.

Self-assessment questions

Complete the following two statements:

95   An instrument which has a class index of 0.5 has an accuracy of _____ _____ .

96   The fiducial value of an instrument is the value to which the _____ _____ .

97   Five instruments have the following class index values:

Instrument no.	1	2	3	4	5
Class index	5	0.1	0.5	2.5	1

List the instruments in order of degree of accuracy with the most accurate first.

98   Complete the following table:

Instrument number	1	2	3	4
class index	0.2	1	5	0.05
fiducial value and f.s.d. value	15 mA	10 A	3000 V	75 mV
permissible error				
percentage error at f.s.d.				
percentage error at three quarters f.s.d.				
percentage error at one half f.s.d.				
percentage error at one quarter f.s.d.				

99   Give two examples each of manufacturing and operating errors which could lead to errors being obtained in the reading obtained from instruments.

100   Explain why an ammeter or voltmeter, which has a fiducial value of f.s.d., is less accurate at its lower scale readings than at its higher scale readings.

After reading the following material, the reader shall:

3.17 Use multimeters for the measurement of *I*, *V* and *R* in d.c. circuits.

3.18 Read multimeter scale values.

3.19 Select correct selector switch setting for given requirements.

3.20 State the symptoms synonymous with battery failure.

From the point of view of both cost and convenience it is desirable to have one instrument that can be used for a variety of measurements. Multirange milliammeters consist of the basic moving-coil instrument capable of being shunted by different values of resistance. The shunt resistances are selected by means of a switch; this provides several different ranges of full-scale current using the one instrument. Similarly the multirange voltmeter consists of the basic moving-coil instrument, which is capable of having various multiplier resistances switched into the circuit to provide different values at full scale deflection. In addition the same type of meter, when used in conjunction with a rectifier, can measure alternating current and p.d. Some meters, known as universal meters or multimeters, combine all these facilities in one instrument, employing only one meter movement which can be switched into a variety of circuits. The various shunts, multipliers, rectifiers and batteries which make up these circuits are all enclosed in one portable case. This widely used class of test instrument is divided into two main groups — analogue and digital.

The analogue type has been in existence since the 1920s. It is used to measure d.c. and a.c. volts, d.c. and a.c. amperes, ohms and decibels. The basic sensitivity of this type of meter is usually 20 000 ohms per volt on d.c. and 5 000 ohms per volt on a.c. When measuring resistance, power is supplied by means of a dry cell, but for other measurements no auxiliary power is required. With the advent of advanced electronics a more sensitive instrument was required. This led to the introduction of an analogue meter capable of making measurements without drawing power from the circuit under test.

Digital multimeters have recently become more widely used, mainly because of the accurate measurements required in modern electronics. However, the analogue instrument with its robustness, cheapness and less servicing requirements is employed in large numbers, particularly for workshop applications.

Figure 120 shows the face of a typical multimeter used in laboratory work. The meter is intended for use horizontally. If the pointer is not on zero, it may be set by adjusting the screw head on the panel.

When measuring current or p.d., the instrument is set to either a.c. or d.c. as appropriate, and to a suitable range before connecting up to the circuit under test. When in doubt the switch should be set to the highest range, and then progressively reduced, if necessary; there is no necessity to disconnect the leads as the switch position is changed. The automatic overload cut-out, if tripped, interrupts the main circuit, and in cases of abnormal overload it provides complete immunity from damage. If the trip operates, the meter leads should be disconnected from the supply, the cut-out reset with the meter horizontal, and the fault rectified before reconnecting the leads. Mechanical shock may also cause the trip to cut out, so the meter should always be handled carefully.

Solutions to self-assessment questions

95   ± 0.5%.

96   Errors of the instrument are referred.

97   Instrument numbers 2, 3, 5, 4, 1.

98

instrument number	1	2	3	4
class index	0.2	1	5	0.05
fiducial and f.s.d. value	15 mA	10 A	3 000 V	75 mV
permissible error $= \dfrac{\text{class index}}{100} \times$ fiducial value	±0.03 mA	± 0.1 A	±150 V	±0.0375 mV
percentage error at f.s.d. $= \dfrac{\text{permissible error}}{\text{f.s.d. value}} \times 100\%$	±0.2%	±1%	±5%	±0.05%
percentage error at three quarters f.s.d. $= \dfrac{4}{3} \times \dfrac{\text{permissible error}}{\text{f.s.d. value}} \times 100\%$	±0.266%	±1.33%	±6.67%	±0.067%
percentage error at one half f.s.d. $= \dfrac{2}{1} \times \dfrac{\text{permissible error}}{\text{f.s.d. value}} \times 100\%$	±0.4%	±2%	±10%	±0.1%
percentage error at one quarter f.s.d. $= \dfrac{4}{1} \times \dfrac{\text{permissible error}}{\text{f.s.d. value}} \times 100\%$	±0.8%	±4%	±20%	±0.2%

99   Manufacturing errors include friction, backlash in gears and incorrect graduation of scales.
Operating errors include parallax error, error due to estimating values.

100   If permissible error is based on f.s.d. value, then at any scale point lower than f.s.d., the percentage error increases as illustrated in Question 98 above.

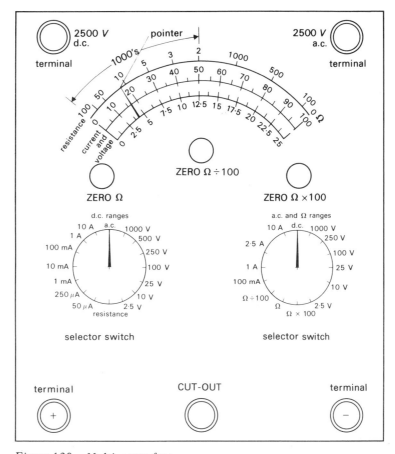

Figure 120   *Multi-meter face*

Measurement of current and p.d. in d.c. circuits

When measuring current and p.d. in d.c. circuits, the right hand switch is set to the d.c. position, and the left hand switch to the range required. To measure the current, the instrument is connected in series with the circuit under test. Before breaking into a circuit to make current measurements, it is essential to ensure that the circuit is 'dead'.

To measure voltage, the instrument is connected across the source of p.d. to be measured. If the p.d. is unknown, the instrument is set to its highest range, connected across the source and, if below 1 000 V, the ranges are decreased step by step until the most suitable range has been selected. If the p.d. should exceed 1 000 V, the instrument should be set at 1 000 V as described above, but the negative lead should be connected to the 2 500 V terminal. Great care must be exercised when making connections to a live circuit; if at all possible the circuit should be made dead before making a connection.

Resistance measurement

Resistance tests should never be carried out on components which are already carrying current. Before making resistance tests the meter pointer should be adjusted to zero in the following sequence:

1   Set the left hand switch at 'Resistance'.
2   Join the leads together.
3   On the Ω range, adjust to zero by means of the knob marked 'ZERO Ω'.
4   On the $\Omega \div 100$ range, adjust to zero by means of the knob marked 'ZERO $\Omega \div 100$'.
5   On the $\Omega \times 100$ range, adjust to zero by means of the knob marked 'ZERO $\Omega \times 100$'.

To test a resistance, the left hand switch is set to resistance, and the right hand switch to the range required, the leads being connected across the unknown component. Resistance is read directly on the Ω range, but scale readings should be divided or multiplied by 100 on the other two ranges. If on joining the leads together it is impossible to obtain a zero ohms setting, or if the pointer position will not remain constant, but falls steadily, the internal battery or cell should be replaced.

Self-assessment questions

Complete the following six statements:

101 An instrument that can measure resistance and current and p.d. in a.c. or d.c. circuits is called a_____.

102 When using a multimeter to measure p.d., the correct circuit selected converts the instrument to a_____ .

103 When using a multimeter to measure current, the correct circuit selected converts the instrument to an_____.

104 When using a multimeter to measure p.d. or current, and assuming the quantity to be measured is within the instrument's ranges, always start measurements on the range which is _____ .

105 If, on trying to measure the resistance of a component, it is found that the pointer position does not remain constant, but falls steadily, this is an indication that the_____ should be replaced.

106 When measuring the resistance of any device in a d.c. circuit using a multimeter, a terminal of the device must be_____.

107 Figure 121 represents a multimeter face. For the pointer position shown, complete the following table of readings:

scale	reading
0–1 000 V d.c.	
0–250 V d.c.	
0–2.5 V d.c.	
0–10 A d.c.	
0–1 mA d.c.	
0–50 μA d.c.	
Ω	
Ω × 100	
Ω ÷ 100	

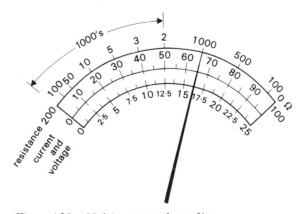

Figure 121   *Multi-meter scale reading*

108   Complete Figure 122 by inserting the selector switch positions so that the multimeter reads the required scales.

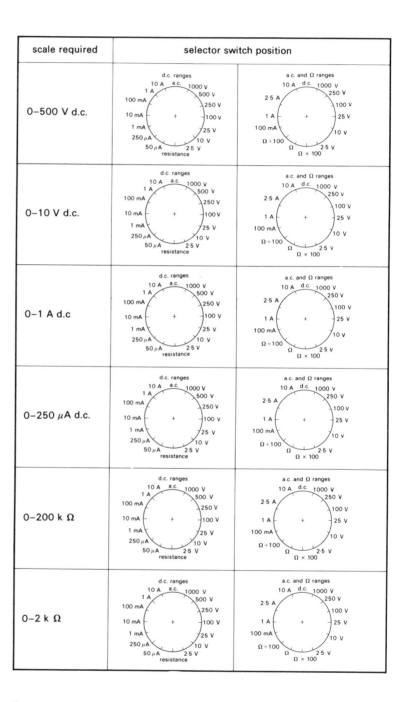

Figure 122   *Multi-meter selector switch position*

109   For the circuit in Figure 123, indicate on the figure, the position of a multimeter which is to measure:

(*a*) the terminal p.d.,

(*b*) the load current,

(*c*) the potential difference across resistor R_3,

(*d*) the current flowing through resistor R_2,

(*e*) the value of the resistance of R_1, stating what precaution must be taken with regard to the d.c. supply.

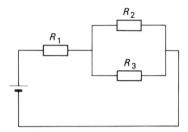

Figure 123   *Multi-meter positions*

Solutions to self-assessment questions

101   Multimeter.

102   Voltmeter.

103   Ammeter.

104   Highest.

105   Batteries or cells.

106   Open circuited.

107

scale	reading
0–1 000 V d.c.	650 V
0–250 V d.c.	162.5 V
0–2.5 V d.c.	1.625 V
0–10 A d.c.	6.5 A
0–1 mA d.c.	0.65 mA
0–50 μA d.c.	32.5 μA
Ω	1 000 Ω
$\Omega \times 100$	100 kΩ
$\Omega \div 100$	10 Ω

Solutions to self-assessment questions

108   See Figure 124.

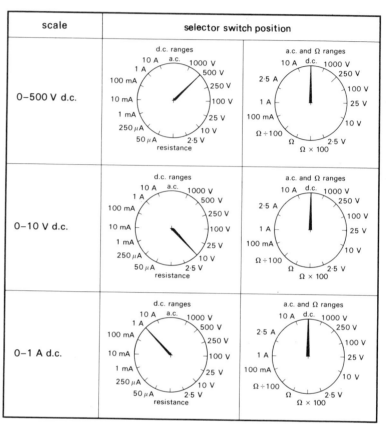

Figure 124   *Solution to self-assessment question 108*

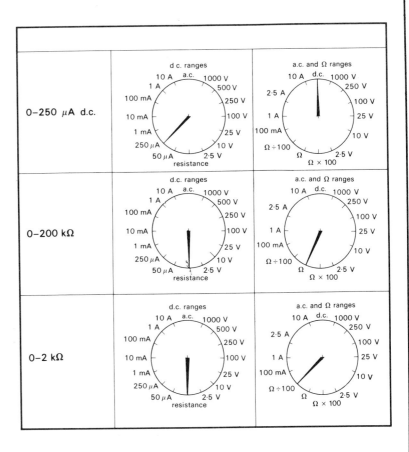

	d.c. ranges	a.c. and Ω ranges
0–250 μA d.c.		
0–200 kΩ		
0–2 kΩ		

109 See Figure 125.

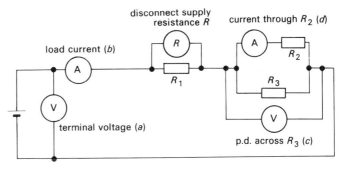

Figure 125 *Solution to self-assessment question 109*

Topic area: Electromagnetic induction

Section 1 Principles of electromagnetic induction

After reading the following material, the reader shall:

1 Know the laws of electromagnetic induction and apply them to the simple generator.

1.1 Describe a simple experiment to illustrate electromagnetic induction.

1.2 State that the induced e.m.f. is proportional to the rate of change of flux linkage.

1.3 Determine the direction of an induced e.m.f. for a conductor moving in a magnetic field.

1.4 State and illustrate Fleming's right hand rule.

1.5 Describe the action of a simple generator.

1.6 Sketch a simple generator.

1.7 Describe the output as an alternating e.m.f. capable of producing an alternating current.

1.8 Explain the difference between single and three phase supplies.

It has already been shown that when an electric current passes through a conductor, a magnetic flux is produced around the conductor. This effect was discovered in 1820 and raised the question: can a magnetic flux produce an electric current? The answer to this question was published in England and America in 1832. Two scientists, Faraday in England and Henry in America found that a magnetic flux could produce an electric current in a conductor. The discovery is credited to Faraday and the phenomenon is called *electromagnetic induction.*

Figure 126 shows one method of demonstrating electromagnetic induction; the arrangement is very similar to the one used by Faraday in 1831. The apparatus consists of a coil (1) wound on an iron ring and connected to a battery. The current in this coil is controlled by a switch. A second coil is wound onto the ring, but in a position different from that of the first coil. The second coil is connected to a moving coil galvanometer which indicates the magnitude of any current flowing in the coil (2).

When the switch is closed, current flows in coil (1); at the same time the needle on the galvanometer deflects and returns to zero, indicating that a current has also passed through coil (2). If the switch is now

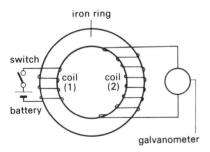

Figure 126 *Demonstration of an induced current*

opened, the galvanometer needle deflects and returns to the zero position, again indicating that a current has passed through coil (2). What has happened to cause the current to flow in coil (2)?

The current flow in coil (1) causes a magnetic flux in the iron ring. When the switch is closed the magnetic flux in the iron ring changes from near zero to a maximum value. It is during this time taken for the flux to reach a maximum value that a current flow is indicated by the galvanometer connected to the coil (2). When the flux reaches a maximum value and remains *constant*, the current flow in coil (2) is *zero*. The *changes in the flux linkages* in the iron ring produce an *induced electro-motive force* in coil (2), which causes a current to flow through the galvanometer.

When the current in coil (1) is switched off, the flux in the iron ring *changes* from a maximum value to a near zero value. During the time that the flux is changing, an e.m.f. is again induced in coil (2), but this time acting in the opposite direction. This induced e.m.f. causes a current to flow through the galvanometer. Note that, *only when the flux is changing is there an induced e.m.f.* This method of producing an induced e.m.f. is used in the transformer, which is a device used to raise or lower voltages. Transformers will be discussed later in this topic area.

Continuing his research, Faraday found that when a permanent magnet was moved near or in a coil, a galvanometer connected to the coil indicated a current flow. Figure 127 shows an arrangement similar to that used by Faraday. He found that the direction of the current flow was dependent upon the direction in which the magnet was moving (i.e. towards or away from the coil), and on whether a north-seeking or a south-seeking pole was nearer to the end of the coil.

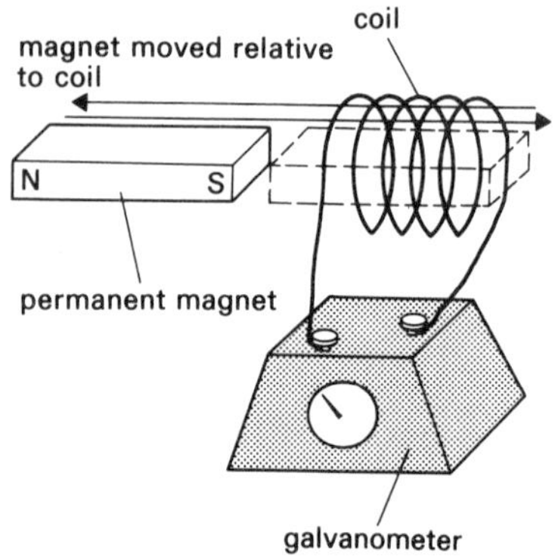

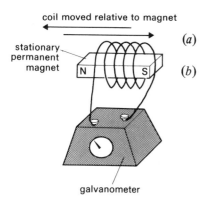

Figure 127 *Simple demonstration of electromagnetic induction*

Faraday also discovered that the magnitude of the induced e.m.f. in the coil is determined by:

(a) the rate at which the magnet is moved towards or away from the coil (i.e. the rate of change of flux linkages in the coil),

(b) the number of turns of conductor in the coil.

The method of inducing an e.m.f. into stationary conductors by a moving magnet is the principle on which the large electrical generators in power stations are designed. As it is relatively easy to insulate stationary conductors, the use of such conductors is a very suitable arrangement in which high e.m.f.s may safely be induced.

coil moved relative to magnet

stationary
permanent
magnet

galvanometer

Figure 128 *Induced electromotive force, coil moving*

An alternative method of producing an induced e.m.f. is to keep the magnet stationary and to move the coil as shown in Figure 128. It is found that the faster the conductor cuts through the field, the more turns there are, and the stronger the magnetic field, the greater is the induced e.m.f., and the greater the current flow. Also, both the polarity of the induced e.m.f., and the direction of current flow, can be reversed by reversing the directions of movement of the conductor.

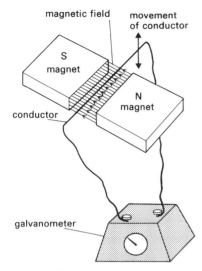

Figure 129 *Conductor moving at right angles to magnetic field*

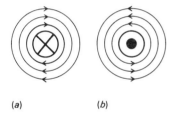

(a)   (b)

Figure 130 *Convention for current flow and magnetic field*

Figure 129 represents a conductor which is free to move upwards or downwards. It is placed so that it moves at right angles to the magnetic field of a permanent magnet. When the conductor is moved, an e.m.f. is induced in the conductor. If the moving conductor forms part of a closed circuit, a current flows in the circuit. This current sets up a magnetic field around the moving conductor, which interacts with the magnetic field of the permanent magnet in such a way as to produce a force on the conductor. This force opposes the force which produces the motion. Using this information a scientist called Lenz devised a law which determines the direction of the induced e.m.f. in the moving conductor. The law is known as Lenz's law and in generalized form states that: *the direction of an induced e.m.f. is such that it tends to set up a flow of current, which in turn causes a force opposing the motion which is generating the e.m.f.*

Note that the direction of current flow is assumed to be from a positive to a negative potential. This direction, known as conventional current flow, is used unless otherwise stated. Figure 130 shows the convention which us used to show the direction of the current flow and the magnetic field around a conductor. This convention was discussed in the previous topic area.

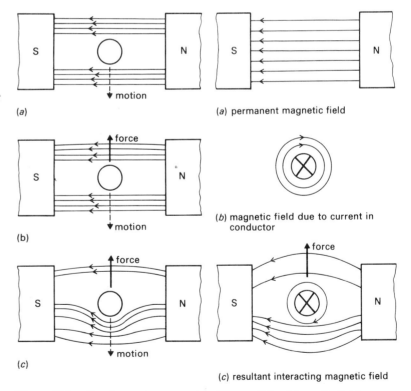

Figure 131 *Conductor in magnetic field*

Figure 132 *Magnetic fields*

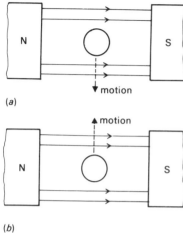

(a)

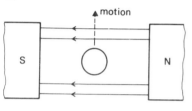

(b)

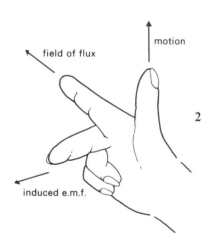

(c)

Figure 133   *Self-assessment question 1*

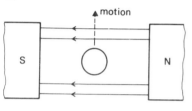

Figure 134   *Fleming's right-hand rule*

Figure 131 (*a*) represents a conductor moving downwards in the magnetic field of a permanent magnet. From Lenz's law, the force on the conductor due to the interaction of the permanent magnetic field and that due to the current in the conductor is upwards, as shown in Figure 131 (*b*). For the force on the conductor to act in an upwards direction, the resulting magnetic field below the conductor must be stronger than that above the conductor, as shown in Figure 131 (*c*). In order to obtain the resultant magnetic field shown in Figure 131 (*c*), the current flow must be as shown in Figure 130 (*a*). The two interacting fields are shown separately in Figure 132 (*a*) and (*b*), and the resultant interacting field is shown in Figure 132 (*c*). Thus the direction of the induced e.m.f. must be such that it causes a current flow which is directed inwards as shown in Figure 132 (*c*).

Self-assessment question

1   Using the procedure described above and the conventions shown in Figure 130, determine the direction of the induced e.m.f. for the motion of the conductor shown in Figure 133 (*a*), (*b*) and (*c*).

An alternative method of determining the direction of the induced e.m.f. is to use Fleming's right hand rule, which is illustrated in Figure 134 and states that:

When the thumb and the first two fingers of the right hand are held at right angles to each other in such a way that the first finger points in the direction of the magnetic field, and the thumb points in the direction of motion, or relative motion, of the conductor, then the second finger points in the direction of the induced e.m.f.

Self-assessment question

2   Use Fleming's right hand rule on the problems shown in Figure 133 (*a*), (*b*) and (*c*) to determine the direction of the induced e.m.f.

The key to understanding the work of Faraday, Henry and Lenz on electromagnetic induction is to appreciate that the *magnetic flux has to be changing before an e.m.f. is induced in the conductor*. The magnitude of the induced e.m.f. is determined by the number of lines of magnetic force or flux which interlink or 'cut' conductors in unit time, so that for a fixed number of conductors:

$$\text{e.m.f.} = \frac{\text{flux change}}{\text{time of flux change}}$$

The results of the various experiments to investigate the effects of electromagnetic induction may be summarized as follows:

(i) the value of the induced e.m.f. in a conductor is proportional to the rate at which the flux linkages change.

(ii) the direction of the induced e.m.f. is such that it tends to set up a flow of current, which in turn causes a force opposing the motion which is generating the e.m.f.

An industrial application of the principle of electromagnetic induction is the generator. The generator is an efficient device which converts available energy into the desired electrical energy.

The simple generator

Figure 135 shows a simplified arrangement of a generator. A coil, represented by a single conductor is mounted on a shaft so that when the shaft is rotated, the coil also rotates between the poles of a magnetic field. The magnetic field is designed to ensure an even distribution of magnetic flux in the air gap between the poles and the conductors. Both ends of the coil are connected to conducting rings, known as *slip rings*, mounted on the shaft. The slip rings rotate with the shaft. Each slip ring is connected to the external circuit by means of a block of carbon called a *carbon brush*. The carbon brushes are stationary, and contact with the slip rings is maintained by springs located so that they exert a force which pushes the brushes onto the slip ring. When the coil rotates, a sliding contact is maintained between the slip rings and the brushes. The carbon brushes are conductors, and conduct the current generated in the coil to the external circuit. The external circuit consists of a load resistor, a voltmeter which measures the p.d. across the resistor and an ammeter which measures the current flow in the resistor at any instant in time. The basic principle on which the generator works is that, as the conductors cut through the magnetic field, an e.m.f. is generated which causes a current to flow through the loop, slip rings, brushes and load. The magnitude and direction of the induced e.m.f., and therefore the magnitude and direction of the current, depends upon the position of the loop in relation to the magnetic field.

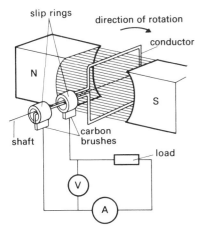

Figure 135 *Simple generator (rotating coil)*

To illustrate the principle of the generator, consider Figure 138. The generator is shown as a coil, represented by two conductors, one plain and the other hatched, in between the poles of a uniform magnetic field. The direction of the magnetic field is from the north-seeking pole towards the south-seeking pole. The two conductors are rotated in a clockwise direction, starting from a zero, horizontal position as shown in Figure 138 (*a*). When the coil is rotating, and at the instant in time represented by Figure 138 (*a*) the conductors are moving in a direction which is parallel to the direction of the lines of force in the magnetic field. At this instant, no lines of magnetic force are being 'cut' or are 'cutting conductors'. Thus the induced e.m.f. is zero, and hence no current is flowing. This value is shown on the graph of current against angular position in Figure 138 (*a*).

As the loop rotates from position 1 to position 3, as shown in Figure 138 (*b*), the conductors are cutting through more and more lines of force. At position 3, that is at 90 degrees, the conductors are cutting through a maximum number of lines of force. Therefore, between zero and 90 degrees the e.m.f. generated, and thus the current in the conductors, increases from zero to a maximum. Note that, from position 1 to position 3, the hatched conductor cuts down through the field, and the plain conductor cuts up through the field. By applying Lenz's law or Fleming's right hand rule it is seen that the resulting e.m.f.s and the currents in both conductors are in series, and the resulting voltage and current flow is the sum of the two individual values.

Solutions to self-assessment questions

1   The direction of the induced e.m.f. is as shown in Figure 136 (*a*), (*b*) and (*c*).

2   The direction of the induced e.m.f. is as shown in Figure 137.

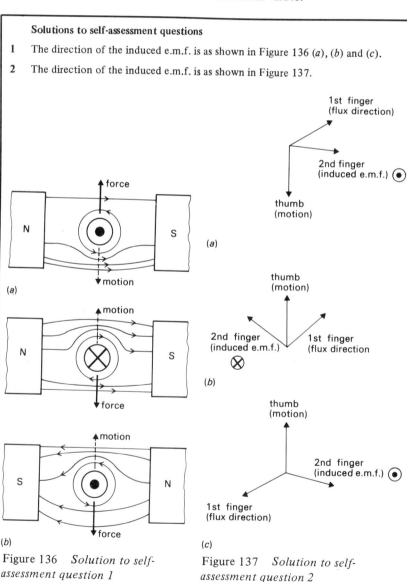

Figure 136   *Solution to self-assessment question 1*

Figure 137   *Solution to self-assessment question 2*

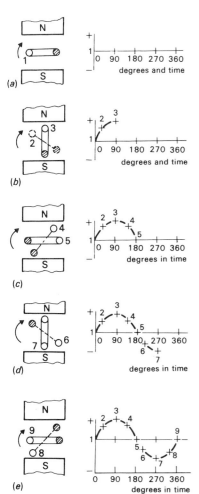

Figure 138 *Generation of alternating voltage*

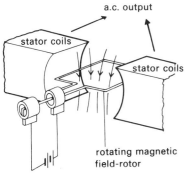

Figure 139 *Rotating field a.c. generator*

As the conductors continue to rotate to position 5, as shown in Figure 138 (*c*), the conductors cut through less and less lines of force, until at position 5 they are moving parallel to the magnetic field and no longer cut through any lines of force. Thus the generated e.m.f. and the current decreases from a maximum at position 3 to a minumum at position 5.

As the loop continues to rotate from position 5 to position 7 as shown in Figure 138 (*d*), the direction of the cutting action of the conductors through the magnetic field reverses. The hatched conductor now cuts up through the field, and the plain conductor cuts down through the field. Because of this reversal in direction, both the polarity of the generated e.m.f. and the current flow reverse. Again, this can be confirmed by applying Lenz's law or Fleming's right hand rule. The values of the generated e.m.f. and of the current increase from a minimum value at position 5 to a maximum at position 7.

Figure 138 (*e*) shows the final part of one complete revolution of the conductors, the conductors moving from position 7 to position 9. As the conductors move from 7 to 9, they cut less and less lines of force, and hence the values of the generated e.m.f. and of the current fall from a maximum at point 7 to a minimum at point 9.

Figure 138 (*e*) represents the waveform of the generated e.m.f. and the current flowing through the load. The generated e.m.f. is not unidirectional but is an 'alternating e.m.f.', since it alternates periodically from positive to negative. The current which flows as a result of this alternating e.m.f. is also alternating, and is referred to as *alternating current (a.c.)*.

Figures 127 and 128 illustrate two different methods of demonstrating electromagnetic induction. In Figure 127 the magnet is moved relative to the coil, while in Figure 128 the coil is moved relative to the magnet. In both cases the outcome is the same — a change in the magnetic flux interlinking with a conductor causing an induced e.m.f. in the conductor. Figure 135 illustrates the principle of the simple generator, showing a conductor rotating in a magnetic field, i.e. a simple extension of the principle illustrated in Figure 128. *This principle of rotating the coil is used only for a.c. generators of small power rating.*

The alternative method of generating an a.c. waveform is in effect to rotate the magnetic field and to keep the coil stationary, i.e. to make use of the principle illustrated in Figure 127. Figure 139 illustrates an a.c. generator in which the coil is fixed and the magnet rotates. The fixed coils are called the *stator*, and the rotating magnet (usually an electromagnet) is called the *rotor*. The advantage of having a stationary coil is that the generated e.m.f. can be connected directly to the load through fixed connectors. It is easier to insulate the fixed connections than it is to insulate the slip rings of the rotating coil type of a.c.

generator. High voltage a.c. generators are usually of the rotating magnetic field type.

It is usual to generate three phase 50 Hz alternating e.m.f. in power stations. If an a.c. generator consists of a single set of fixed coils, then the resultant waveform is as shown in Figure 138 (*e*). In a three phase a.c. generator three sets of fixed coils are used. The three sets of fixed coils are so spaced that the e.m.f. induced in any one is displaced by 120° from the other two. Figure 140 (*a*) represents a three phase a.c. generator, and Figure 140 (*b*) shows the waveform obtained from such a generator.

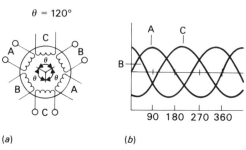

(*a*)   (*b*)

Figure 140   *Three-phase a.c. generator*

The three phase system for generation and distribution of large power supplies has been adopted as standard throughout the world for reasons which include:

1   A three phase distribution system requires less weight of conductor than a single phase system provided that,
    (*a*) the same quantity of power is supplied,
    (*b*) the power supplied is at the same p.d.,
    (*c*) the load is the same distance from the source of power.

2   A three phase electric motor operating at a given rate, and on a given p.d., is cheaper to purchase, is more efficient and has a smaller frame size than a single phase electric motor operating under the same conditions.

The power produced in a power station at 11 kV or 22 kV has to be distributed to industrial, commercial and domestic users. To accomplish this and to provide each class of consumer with p.d. to match requirements, the output p.d. of each power station is raised to the same level. This enables the output of all power stations to be fed into a distribution system. Figure 141 shows a layout of a transmission and distribution system.

In Figure 141 the power station (A) has an output which is three phase 50 Hz 22 kV. This three wire supply is fed to a transformer (B) which

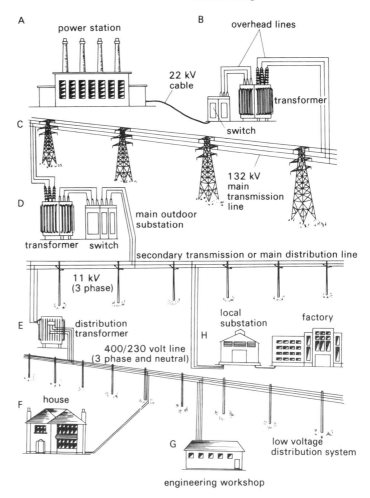

Figure 141 *Layout of transmission and distribution system*

raises the p.d. to 132 kV. This high p.d. enables the transmission of high powers over long distances at low currents, with subsequent savings on conductor sizes. The main transmission lines are supported by pylons (C). The 132 kV supply is reduced by the transformer (D) to 11 kV for transmission around a local area on overhead lines secured to poles. The 11 kV supply feeds the transformer (E) which reduces the p.d. to a value which is suitable for domestic use as shown at (F). The power requirements of small industrial and commercial consumers (G) are usually supplied by the low p.d. distribution system. Larger industrial concerns (H) may require supplies at 11 kV or 6.6 kV in addition to a low p.d. system. Local substations are used to enable these demands to be met.

Self-assessment questions

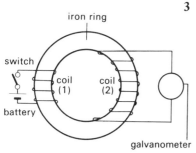

Figure 142   *Demonstration of an induced current*

3   Mark each of the following statements as either true or false:
   (i)   Figure 142 shows one method of demonstrating electromagnetic induction; there is a flow of current in the galvanometer when:
      (*a*)   the current in the coil (1) is constant.

                                                                 TRUE/FALSE

      (*b*)   the flux linkages in coil (2) are changing.

                                                                 TRUE/FALSE

      (*c*)   the flux is constant.

                                                                 TRUE/FALSE

   (ii)   If the number of turns in coil (2), shown in Figure 142, is doubled, then the deflection of the galvanometer, a fraction of a second after the switch was closed, would be halved.

                                                                 TRUE/FALSE

   (iii)   In the simplified generator shown in Figure 137
      (*a*)   the flux density between the poles of the permanent magnet is constant.

                                                                 TRUE/FALSE

      (*b*)   the coil sides interlink with the lines of magnetic force at the same rate all the time.

                                                                 TRUE/FALSE

   (iv)   If the flux density of the field shown in Figure 137 is halved, the maximum value of the current in the resistor is halved, other factors being maintained constant.

                                                                 TRUE/FALSE

   (v)   The waveform of the current passing through the resistor shown in Figure 137 for one complete revolution of the conductor is in the form of an alternating current.

                                                                 TRUE/FALSE

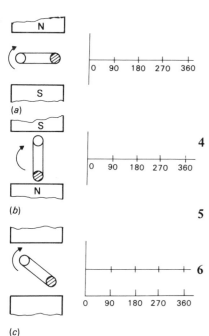

(a)

(b)

(c)

Figure 143   *Self-assessment question 4*

4   Figure 143 (*a*), (*b*) and (*c*) show two conductors rotating in a clockwise direction in a magnetic field. Assuming that in each case shown, the conductor is at the zero degree position, sketch the resulting e.m.f. waveform on the graph axes supplied.

5   Using the sign convention shown in Figure 130 to illustrate the answers, determine the direction of the induced e.m.f. for the motion of the conductor shown in Figure 143 (*a*) and (*b*).

6   Complete the following statements by adding the missing word or words:
   (i)   An induced e.m.f. is proportional to the rate of change of _____.
   (ii)   Most industrial a.c. generators work on the rotating _____ principle.
   (iii)   In the three phase a.c. generator there are three single phase windings spaced at _____ degrees to each other.

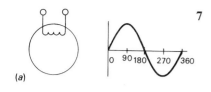

(a)

7   Figure 144 (*a*) shows the waveform for a single phase a.c. generator. Complete Figure 144 (*b*) to illustrate the difference in waveform between a single phase and a three phase supply.

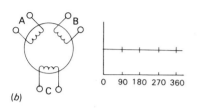

(b)

Figure 144   *Self-assessment question 7*

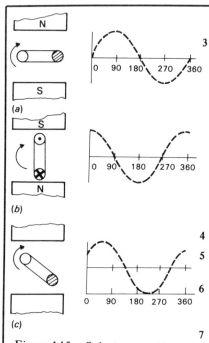

Figure 145   *Solution to self-assessment question 4*

3   (i)   (a)   FALSE.
            (b)   TRUE.
            (c)   FALSE.
            The flux must be changing to produce a current flow in the galvanometer.
      (ii)   FALSE – the deflection is doubled because, as the number of conductors is increased, so is the number of interlinkages between the conductors and the flux.
      (iii) (a)   TRUE.
            (b)   FALSE – as the conductor rotates, the interlinkages vary from a maximum, when the conductor is moving at right angles to the direction of lines of magnetic force, to a minimum when the conductor is moving parallel to the lines of magnetic force.
      (iv)   TRUE.
      (v)   TRUE.

4   The resulting waveforms are as shown in Figure 145.

5   (a)   e.m.f. produces no current flow.
      (b)   e.m.f. produces current flow as shown in Figure 145 (b).

6   (i)   Flux linkages.
      (ii)   Magnetic field.
      (iii) $120°$.

7   The resulting waveform is shown in Figure 140.

Section 2
Characteristics of alternating current

After reading the following material, the reader shall:

2   Appreciate the characteristics of alternating current.
2.1 State that the mains supply is of sinewave form.
2.2 State that the standard commercial frequency in Great Britain is 50 Hz.
2.3 Define peak value.
2.4 Define frequency.
2.5 Define peak to peak value.
2.6 Calculate root mean square values from peak values.
2.7 Calculate peak values from root mean square values.

Figure 138 (*e*) shows one complete cycle of the generated e.m.f. In producing the waveform shown in Figure 138 (*e*) the loop has rotated through 360°, i.e. one complete revolution. One complete cycle of generated e.m.f. is produced for each revolution the loop makes. Thus, if the loop rotates at a speed of 50 revolutions per second, the generated e.m.f. completes 50 cycles per second. *The number of cycles completed in unit time is known as the frequency.* Thus, if the generated e.m.f. completes 50 cycles per second, then it is said to have a frequency of 50 cycles per second. The preferred unit for frequency is the hertz, where

1 hertz = one cycle per second

The symbol used for frequency is f, and the abbreviation of hertz is Hz.

The frequency of the supply in the UK is 50 Hz ± 1%, and all a.c. equipment manufactured for use in the UK is designed to operate within these limits of frequency. If the frequency changes, the performance of a.c. equipment changes, e.g. the speed of electric motors is sensitive to frequency changes.

The waveform from an a.c. supply is usually assumed to be represented by a sine curve. This is not always true, but for the purposes of this unit it will be assumed that mains a.c. supply has a sine waveform. Figure 146 represents one cycle of an a.c. supply in the form of a sine curve.

Some of the terminology associated with the waveform of an a.c. supply is explained in the following paragraphs. The term one cycle has already been explained. During one complete cycle there are *two maximum or peak values*, one for the positive half cycle and the other for the negative half cycle. The difference between the peak positive

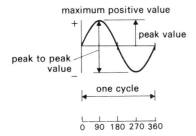

Figure 146   *One cycle of a.c.*

value and the peak negative value is called the peak to peak value of a sine wave. This value is twice the maximum or peak value of the sine wave.

The most common method of expressing the value of a current and p.d. in a.c. circuits is to state the *root mean square (r.m.s.)* value. To understand the concept of root mean square values, consider a current having a sine waveform flowing through a resistor which has a resistance of R ohms. The value of the current flowing through the resistor varies continuously, but when any current flows through a resistor, the resistor heats up. The heating effect of a direct current can be compared with the heating effect of an alternating current. It is found that the heating effect of a direct current passing through a resistor of resistance R ohms for t seconds is given by the product of $I^2 R t$, where I is the current in amperes.

When the same resistor is used for the same time with a.c., it can be shown that,
$$I^2 R t = \tfrac{1}{2} I_{max}^2 R t$$

It follows that the effective value of a.c. of I amperes is:
$$I = \sqrt{\tfrac{1}{2}} I_{max}$$
i.e. $I = 0.707 I_{max}$

This effective value is known as the root mean square value, and is widely used in calculations.
$$I_{r.m.s.} = 0.707 I_{max}$$
and $V_{r.m.s.} = 0.707 V_{max}$

Example 1
An alternating e.m.f. having a sinusoidal waveform reaches a maximum value of 200 V. Calculate the r.m.s. value of the e.m.f.

For a sinusoidal waveform, the r.m.s. value is
$$V_{r.m.s.} = 0.707 V_{max}$$
therefore $V_{r.m.s.} = 0.707 \times 200$
$$= 141.4 \text{ V}$$

Example 2
An alternating e.m.f has a sinusoidal waveform of peak value of 283 V. What is the r.m.s. value of the e.m.f.?
$$V_{r.m.s.} = 0.707 V_{max}$$
therefore $V_{r.m.s.} = 0.707 \times 283$
$$V_{r.m.s.} = 200 \text{ V}$$

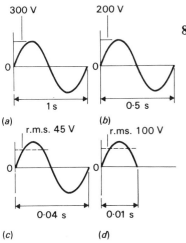

(a)

(b)

(c)

(d)

Figure 147   *Self-assessment question 9*

Self-assessment questions

8   Mark each of the following statements as either true or false.

(i)   The waveform of the mains a.c. supply is called a sine wave.

TRUE/FALSE

(ii)   In S.I. the preferred unit of frequency is the number of cycles completed per minute.

TRUE/FALSE

(iii)   The peak value is the maximum positive or negative value in a cycle of a.c.

TRUE/FALSE

(iv)   In one complete cycle of a.c. there are two peak values.

TRUE/FALSE

(v)   The difference between the peak positive value and the peak negative value is called the peak to peak value.

TRUE/FALSE

(vi)   The standard frequency used in Great Britain is 500 Hz.

TRUE/FALSE

9   Figure 147 shows the e.m.f. waveform generated by a loop rotating at different speeds. The time taken to generate each waveform is shown under each waveform. Calculate the frequency for each generated e.m.f.

10   On the waveform shown in Figure 147 (*a*) indicate the peak positive, peak negative and peak to peak values.

11   Calculate the r.m.s. values for the waveforms shown in Figures 147 (*a*) and (*b*).

12   Calculate the peak value for the waveform shown in Figures 147 (*c*) and (*d*).

Solutions to self-assessment questions

8 (i)   TRUE.

   (ii) FALSE – the preferred unit of frequency is the number of cycles completed per second.

   (iii) TRUE.

   (iv) TRUE.

   (v)   TRUE.

   (vi) FALSE – standard commercial frequency is 50 Hz.

9 Frequency is the number of cycles completed per second.

   (a) Frequency $= \dfrac{1}{1} = 1$ Hz.

   (b) Frequency $= \dfrac{1}{0.5} = 2$ Hz.

   (c) Frequency $= \dfrac{1}{0.04} = 25$ Hz.

   (d) Frequency $= \dfrac{1}{0.02} = 50$ Hz.

10 The values are as shown in Figure 148.

11 (a) r.m.s. value = $0.707 \times$ peak value,

                       = $0.707 \times 300 = 212.1$ V

    (b) r.m.s. value = $0.707 \times 200 = 141.4$ V

12 (a) Figure 147 (c) peak value $= \dfrac{\text{r.m.s. value}}{0.707}$

                                    $= \dfrac{45}{0.707} = 63.65$ V

    (b) Figure 147 (d) peak value $= \dfrac{100}{0.707} = 141.44$ V

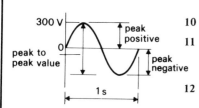

Figure 148   *Solution to self-assessment question 9*

Section 3
The transformer

After reading the following material, the reader shall:

3   Understand the function of a transformer.

3.1   Describe the phenomenon of mutual induction.

3.2   Describe the action of a transformer.

3.3   Draw the correct symbol for a transformer.

3.4   Solve simple problems on the transformer, e.g. voltage, current and turns ratio assuming 100% efficiency.

As mentioned earlier, the p.d. of an a.c. supply can be raised or lowered by means of a transformer. To transmit a.c. power at a high p.d./low current level, the energy generated is connected to a transformer; the transformer raises the p.d. and since power depends on both p.d. and current, a higher p.d. means that the same amount of power requires a lower current. At the load end of the transmission line, another transformer reduces the p.d. down to the required value. In addition to their use in transmission systems, transformers are used in many types of electronic equipment to raise and lower the p.d. in a.c. circuits.

A single phase, double wound transformer consists of two coils, which are electrically insulated from each other, mounted on a ferrous core in a manner similar to that which Faraday used in his experiment on electromagnetic induction. In diagrammatic form this arrangement is shown in Figure 149 (*a*). The symbol used to show a transformer in circuit diagrams is shown in Figure 149 (*b*). The coil to which the supply is applied is called the primary coil, and the coil in which the p.d. is induced is called the secondary coil.

Figure 150 shows a primary coil A and a secondary coil B. When the primary coil A carries an alternating, or changing, current, the magnetic field shown in Figure 150 (*a*) alternately increases and collapses. If the primary coil A is placed beside the secondary coil B, then the changing magnetic field due to the alternating current flowing in coil A will induce an e.m.f. in coil B. The e.m.f. induced into the secondary coil B is known as the e.m.f. of *mutual inductance*. Note that there is no direct electrical connection between the two coils, only a magnetic connection, i.e. a magnetic linkage. The magnetic linkage is improved when both coils are mounted on a ferrous core.

The alternating current which is flowing in the primary coil, i.e. the magnetizing current, produces an alternating flux in the iron core. This alternating flux induces an e.m.f. in the secondary coil. It is found that

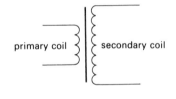

primary coil   secondary coil

(*a*)

(*b*)

Figure 149   *Transformer*

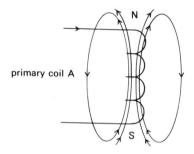

(a)

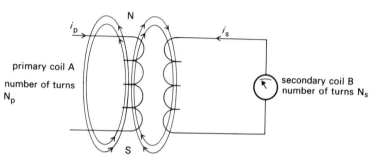

(b)

Figure 150   *e.m.f. due to mutual inductance*

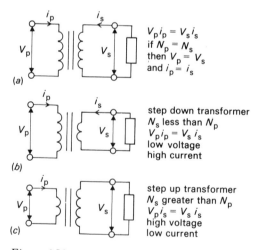

Figure 151   *Types of transformer*

the e.m.f. induced in the secondary coil depends on the ratio of the number of turns in the secondary coil to the number of turns in the primary coil. For example, if there are 500 turns in the secondary coil and only 100 turns in the primary coil, the e.m.f. induced in the secondary coil is 5 times the p.d. applied across the primary coil (500/100 = 5). Since there are more turns on the secondary coil than there are on the primary coil, this type of transformer is called a 'step-up transformer'. If, on the other hand, the primary coil has 10 times as many turns as the secondary coil, the e.m.f. induced in the secondary coil is one tenth of the p.d. applied across the primary coil. When there are fewer turns on the secondary coil than there are on the primary coil, the transformer is called a 'step-down transformer'.

Consider the coils shown in Figure 151. Note that the secondary coil is connected to a load and the p.d. across the load is referred to as V_s.

Let V_p = primary p.d.
$\quad V_s$ = secondary p.d.
and N_p = number of turns on primary coil
$\quad N_s$ = number of turns on secondary coil
and i_p = current in primary coil
$\quad i_s$ = current in secondary coil.

Assuming that there are no losses, i.e. the transformer is 100% efficient, then the primary power must equal the secondary power.
i.e. $V_p i_p = V_s i_s$

and since the induced e.m.f. is proportional to the number of turns, as it is assumed that there are no losses,

$$\frac{V_p}{V_s} = \frac{N_p}{N_s}$$

Example 3
A transformer has a step down ratio of 20:1. Calculate the p.d. across the secondary coil when the primary supply is 240 V. Assume that there are no losses.

For a step down transformer the number of turns on the secondary coil is less than on the primary coil.

$$\frac{V_p}{V_s} = \frac{N_p}{N_s} = \frac{20}{1}$$

Hence $\quad V_s = \dfrac{V_p}{20} = \dfrac{240}{20}$

$\quad V_s = 12$ V

Example 4
A transformer is used to supply 50 V from the 250 V mains. The primary winding contains 1 500 turns. Calculate:

(a)   the number of secondary turns,
(b)   the secondary current when the primary current is 5 A.
       Assume there are no losses.

Consider the transformer in Figure 152.

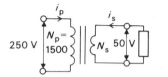

Figure 152   *Example 4*

(a)
$$\frac{V_p}{V_s} = \frac{N_p}{N_s}$$

Transposing, $N_s = N_p \dfrac{V_s}{V_p}$

Using the values, $N_s = 1\,500 \times \dfrac{50}{250}$

$N_s = 300$ turns

(b)   Assuming no losses, $V_p i_p = V_s i_s$

Transposing, $i_s = \dfrac{V_p}{V_s} i_p$

$i_s = \dfrac{250}{50} \times 5$

$i_s = 25$ A

Example 5
The primary winding of a step down transformer takes a current of 6 A at 2 000 V. If the transformer ratio is 20:1, calculate:
(a)   the p.d. across the secondary winding, and
(b)   the secondary current.
       Assume that there are no losses.

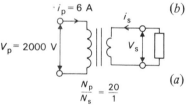

Figure 153   *Example 5*

The arrangement is shown in Figure 153.

(a)   Using, $\dfrac{V_p}{V_s} = \dfrac{N_p}{N_s}$

Transposing, $V_s = V_p \dfrac{N_s}{N_p}$

therefore, $V_s = 2\,000 \times \dfrac{1}{20} = 100$ V

(b)   Using $V_p i_p = V_s i_s$

$i_s = \dfrac{V_p i_p}{V_s}$

therefore, $i_s = \dfrac{2\,000}{100} \times 6$

$i_s = 120$ A

Self-assessment questions

Complete the following statements by adding the missing word or words.

13 (i) The p.d. of an a.c. supply can be raised or lowered by means of a _____ .

(ii) In a transformer, the coil to which the supply is connected is called the _____ coil.

(iii) Assuming that the transformer is 100% efficient, then the product of p.d. and current for the primary and secondary coils is _____ .

(iv) The e.m.f. induced in the secondary coil of a transformer is caused by the phenomenon known as _____ .

(v) A transformer is represented by the symbol _____ .

(vi) In a transformer, the coil in which the e.m.f. is induced is called the _____ coil.

Mark each of the following statements as either true or false.

14 (i) There is a direct electrical connection between the two coils of a transformer.

TRUE/FALSE

(ii) In a step-up transformer, the secondary coil has the larger number of turns.

TRUE/FALSE

(iii) For the transformer shown in Figure 154, the p.d. across the primary coil is less than the e.m.f. induced in the secondary coil.

TRUE/FALSE

(iv) For the transformer shown in Figure 154, assuming no losses, the current flowing in the secondary coil is N_s/N_p times the value of the current flowing in the primary coil.

TRUE/FALSE

(v) For the transformer in Figure 155, the current flowing in the primary coil is greater than the current flowing in the secondary coil.

TRUE/FALSE

(vi) For the transformer in Figure 155, the e.m.f. induced in the secondary coil is N_s/N_p times the value of the p.d. across the primary coil.

TRUE/FALSE

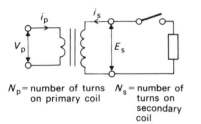

N_p = number of turns on primary coil N_s = number of turns on secondary coil

Figure 154 *Self-assessment question 14*

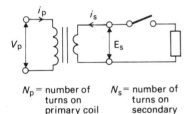

N_p = number of turns on primary coil N_s = number of turns on secondary coil

Figure 155 *Self-assessment question 14*

15 In a single phase double wound transformer the primary coil has 3 000 turns and the secondary coil has 100 turns. Calculate the p.d. across the secondary coil if the p.d. across the primary coil is 240 V.

16 A transformer is used to supply 60 V from the 240 V mains. If the primary winding contains 1 600 turns, calculate:

(*a*) the number of secondary turns,

(*b*) the secondary current when the primary current is 4 A.

17 The primary winding of a step down transformer takes a current of 5 A at 1 600 V. If the transformer ratio is 20:1, calculate the values of the secondary current and the p.d. across the coil.

Solutions to self-assessment questions

13   (i)   Transformer.
     (ii)   Primary.
     (iii)   The same.
     (iv)   Mutual induction.
     (v)   The symbol is as shown in Figure 156.
     (vi)   Secondary.

Figure 156   *Solution to self-assessment question 13*

14   (i)   FALSE. The connection is magnetic.
     (ii)   TRUE.
     (iii)   TRUE.
     (iv)   FALSE – in a step up transformer, the induced e.m.f. is increased by the ratio N_s/N_p; the current is smaller by the same ratio.
     (v)   FALSE – in a step down transformer, the p.d. across the secondary coil is less than the p.d. across the primary coil; therefore the current in the secondary coil is greater than the current flowing in the primary coil.
     (vi)   TRUE.

15   The p.d. across the secondary coil, $V_s = \dfrac{N_s}{N_p} V_p = \dfrac{100}{3\,000} \times 240 = 8 \text{ V}$

16   (a) $\dfrac{V_p}{V_s} = \dfrac{N_p}{N_s}$ $\therefore N_s = N_p \dfrac{V_s}{V_p} = 1\,600 \times \dfrac{60}{240} = 400$

     (b)   Assuming 100% efficiency, $V_p\, i_p = V_s\, i_s$
     $$240 \times 4 = 60 \times i_s$$
     $$\therefore i_s = \frac{240 \times 4}{60} = 16 \text{ A.}$$

17   $V_s = \dfrac{1\,600}{20} = 80 \text{ V}$

     Assuming 100% efficiency, $V_p\, i_p = V_s\, i_s$

     $\therefore i_s = \dfrac{V_p\, i_p}{V_s} = \dfrac{1\,600 \times 5}{80} = 100 \text{ A}$

Topic area: Statics

Section 1   Effects of forces on materials

After reading the following material, the reader shall:

1   Appreciate the effect of forces on materials used in engineering and the need for care in design and material selection.

1.1   Recognize tensile, compressive and shear forces.

1.2   Identify tensile and compressive forces as normal forces.

1.3   Identify shear forces as transverse forces.

In order to help to appreciate the effect of forces on materials consider Newton's third law which can be stated as: *to every action there is an equal and opposite reaction.* In simple terms this means that every push (or pull) must be matched and balanced by an equal and opposite push (or pull). It does not matter how the push (or pull) arises. For example,

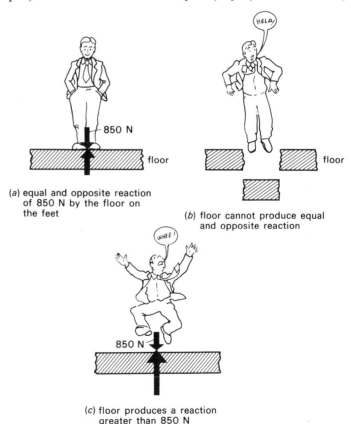

(a) equal and opposite reaction of 850 N by the floor on the feet

(b) floor cannot produce equal and opposite reaction

(c) floor produces a reaction greater than 850 N

Figure 157   *Floor reaction*

if a person has a mass such that, when standing up, he exerts a force of 850 N on the floor, then the soles of the feet push downwards on the floor with a push of 850 N. The floor reacts and pushes upwards with a force of 850 N as illustrated in Figure 157 (*a*). What happens if the floor does not offer this equal and opposite reaction? Two alternatives are possible. If the floor is rotten and cannot provide an equal and opposite reaction, then the person falls through the floor as shown in Figure 157 (*b*). If on the other hand the floor produces a reaction greater than the 850 N, then the person becomes airborne, as illustrated in Figure 157 (*c*). Needless to say, the first alternative is possible, the second would be a miracle! The concept of Newton's third law is not restricted to stationary bodies; for example, if a motor cycle is driven into a wall, the wall may respond by producing enough force to stop the motorcycle at whatever speed it may be going as shown in Figure 158.

Both these examples serve to illustrate Newton's third law. This law does not say anything about how the various forces are generated. The

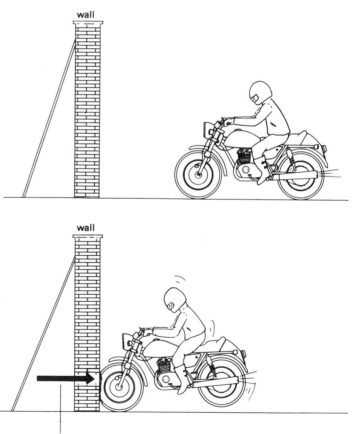

Figure 158   *Newton's third law*

forced produced to stop motor cyclist

push on the floor arises from the action of the earth's gravitation upon the mass of the person, whilst with the motorcycle the forces generated are those needed to decelerate the moving mass, i.e. the application of Newton's second law, which is dealt with in more detail in the next topic area. In this topic area consideration will be given only to bodies in equilibrium.

It is necessary for the reader to be able to recognize three basic types of force which occur in engineering. The three types of force are tensile, compressive and shear.

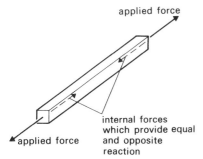

Figure 159 *Material subjected to tensile forces*

Tensile forces

Forces which tend to stretch a material are called tensile forces. A material subjected to tensile forces is said to be in tension. Figure 159 illustrates a material in tension. The rope between two tug-of-war teams is in tension. The rope of a crane lifting an engineering component is in tension. A bolt is put in tension when the nut is tightened.

In Figure 159 the external or applied forces are shown by single arrows pulling away from the material to indicate that the material is being stretched, i.e. it is in tension. The equal and opposite reactions are produced in the material, and are shown by the dotted arrows. How these internal forces are produced will be discussed later. Figure 160

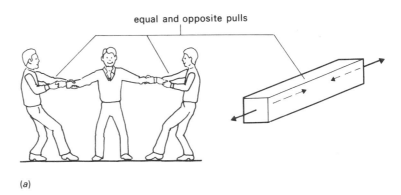

(a)

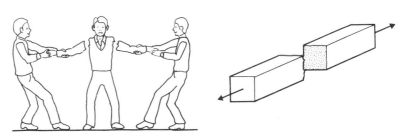

Figure 160 *Effects of applied forces*

(b)

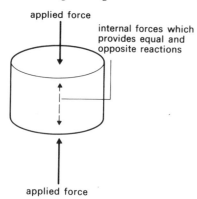

Figure 161   *Material subjected to compressive forces*

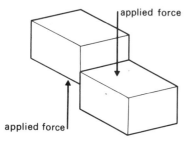

Figure 162   *Material subjected to shearing forces*

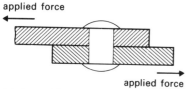

(a) rivet subjected to shear

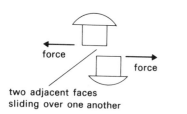

(b) rivet failing in shear

Figure 163   *Rivet in shear*

represents a simple demonstration of the development of internal forces. Figure 160 (*a*) depicts two people pulling on the arms of a third person. The natural reaction of the person in the centre is to pull against the applied forces, that is, to produce an equal and opposite reaction. If the applied forces are increased until the material cannot produce an equal and opposite reaction then the material will fail, as illustrated in Figure 160 (*b*).

Compressive forces
Forces which tend to compress or squeeze a material are called compressive forces. A material subjected to compressive forces is said to be in compression. Figure 161 illustrates a material in compression; note that the internal forces push outwards in this case to produce equal and opposite reactions. The legs of a chair are in compression when someone sits on the chair. A component gripped between the jaws of a vice is subjected to compressive forces.

The lines of action of tensile and compressive forces are such that the forces act normally (i.e. at 90°) to the cross-section of the material. *Thus, tensile and compressive forces are termed normal forces.*

Shear forces
Forces which tend to cause one face of material to slide over an adjacent face (as shown in Figure 162), are called shear forces. Figure 163 (*a*) illustrates a rivet subjected to shear forces. If the rivet were to fail due to the shear forces then Figure 163 (*b*) illustrates the mode of failure. The forces at the blades of a pair of scissors cutting paper, or at the blades of a guillotine shearing machine cutting steel plate, are shearing forces.

The shear forces in Figures 162 and 163 act across the area of the material, i.e. the force acts in a transverse direction. *Thus a shear force is termed a transverse force.*

In practice, a material may be subjected to a combination of tensile, compressive and shearing forces. However, any study of these more complicated conditions must be based on an understanding of the effects of tensile, compressive and shearing forces when they act alone. The examples considered in this book will be restricted to the effects of tensile, compressive and shearing forces acting alone.

Self-assessment questions

1   Answer true or false to each of the following statements:
(i)   Tensile forces always cause an extension in length.
                                                                 TRUE/FALSE
(ii)   Compressive forces always cause a shortening in length.
                                                                 TRUE/FALSE

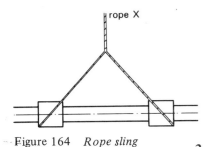

Figure 164 *Rope sling*

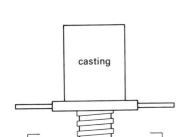

Figure 165 *Shaft supported between centres*

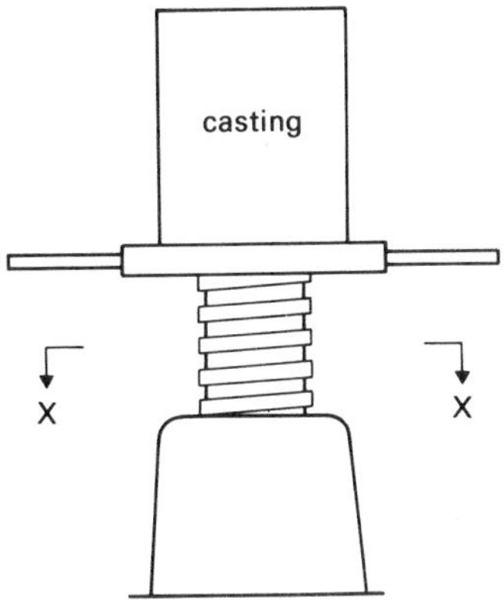

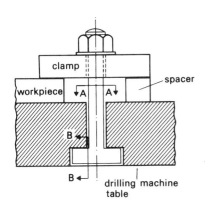

Figure 166 *Screw jack*

(iii) Shearing forces involve cutting actions.

TRUE/FALSE

(iv) 'Normal' forces act at 90° to the cross-section of the material.

TRUE/FALSE

(v) Tensile and compressive forces are called transverse forces.

TRUE/FALSE

(vi) Shearing forces are called normal forces.

TRUE/FALSE

2 Study each of the following statements and associated figures carefully, and then select the word which correctly completes the statement:

(i) Figure 164 shows a shaft being lifted by a rope, X. This rope is subjected to (TENSILE/COMPRESSIVE/SHEAR) forces.

(ii) Figure 165 shows a shaft supported between the centres of a lathe. The shaft is subjected to (TENSILE/COMPRESSIVE/SHEAR) forces.

(iii) Figure 166 shows a screw jack supporting a casting. The screw thread at section XX is subjected to (TENSILE/COMPRESSIVE) forces.

(iv) Figure 167 shows a component clamped to a drilling machine table.

   (*a*) The bolt at section AA is subjected to (TENSILE/COMPRESSIVE/SHEAR) forces.

   (*b*) The bolt at section BB is subjected to (TENSILE/COMPRESSIVE/SHEAR) forces.

(v) Figure 168 shows a transmission joint subjected to forces F as shown.

   (*a*) The bolt A is subjected to (TENSILE/COMPRESSIVE/SHEAR) forces.

   (*b*) The component B is subjected to (TENSILE/COMPRESSIVE/SHEAR) forces.

   (*c*) The component C is subjected to (TENSILE/COMPRESSIVE/SHEAR) forces.

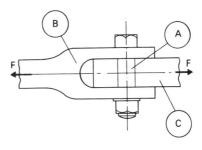

Figure 167 *Clamping of material*

Figure 168 *Transmission joint*

After reading the following material, the reader shall:

1.4   Define stress as the force per unit cross-sectional area.
1.5   Solve simple problems involving direct stress.
1.6   Define strain as change in dimension per unit original dimension.
1.7   Solve simple problems involving strain.

How do materials generate the internal forces to oppose the external applied forces? Human beings and other animals resist mechanical forces by pushing back in an active way, they tense their muscles and push or pull as the situation requires. Engineering materials are passive and cannot push back deliberately. They produce internal resisting forces only in response to externally applied forces which tend to distort the body. Figure 169 illustrates the distortion taking place in a piece of rubber in order to generate sufficient internal forces to support externally applied forces. In Figure 169 (*a*) the rubber can be considered to be unloaded. If a mass *m* is attached to the free end of the piece of rubber then the rubber stretches until it generates internal forces sufficient to support the mass as shown in Figure 169 (*b*).

Figure 170 illustrates a material subjected to compressive forces; in this case the material contracts thus producing the internal forces required to balance the applied external forces. Figure 171 illustrates the forces in the legs of a chair when a person sits on it. The chair legs are in compression due to the forces, and contract until they produce sufficient forces to balance the forces due to the body sitting on the chair.

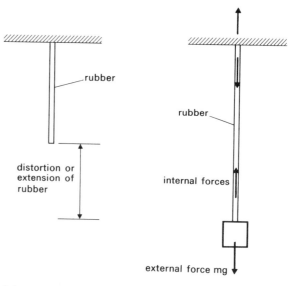

Figure 169   *Material in tension*   (a)   (b)

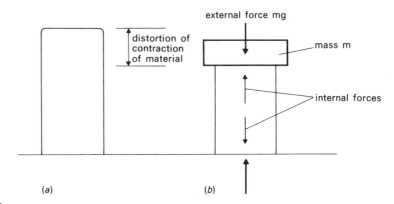

Figure 170   *Material in compression*

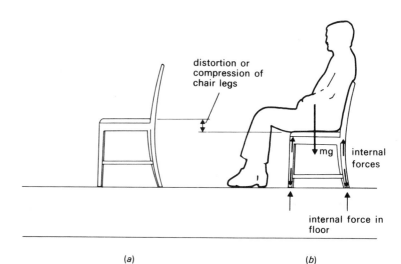

Figure 171   *Forces in chair legs*

Solutions to self-assessment questions

1   (i)   TRUE. Tensile forces always cause an extension in length, although the extension may not always be visible to the human eye.
    (ii)   TRUE. Compressive forces cause a shortening in length.
    (iii) TRUE. Shear forces are transverse forces causing a cutting action.
    (iv)   TRUE.
    (v)   FALSE. Tensile and compressive forces are normal forces.
    (vi)   FALSE. Shearing forces are transverse forces giving a cutting action.

2   (i)   TENSILE.
    (ii)   COMPRESSIVE.
    (iii) COMPRESSIVE.
    (iv)   (a)   TENSILE.
           (b)   SHEAR.
    (v)   (a)   SHEAR.
           (b)   TENSILE.
           (c)   TENSILE.

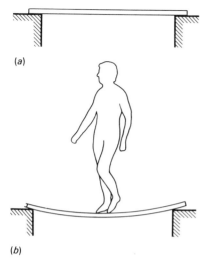

(a)

(b)

Figure 172   *Reaction to a force*

Figure 172 (*a*) represents a plank spanning a gap. When the plank is used as a bridge as shown in Figure 172 (*b*) it deflects and in so doing produces reactions equal and opposite to the applied force due to the person standing on the plank. *In general terms a body in equilibrium is deflected exactly enough to build up forces which just counter the external forces applied to it.*

Stress

In Figures 159 and 161 each internal force is shown as a single arrow. In actual fact the whole of the cross-sectional area of the material at any cross-section within the material helps to produce the internal reactions to the applied forces. In diagrammatic form the force lines are as shown in Figure 173 (*a*). The atoms in a solid are held together by chemical forces or bonds, and when a force is applied to a material each of these individual chemical bonds offers its own individual resisting force. The total of all these individual forces is equal to the applied force, and represents the equal and opposite reaction. *The ratio of internal force and cross-sectional area, that is the force per unit area, is given the name stress.*

Figure 173 represents a bar subjected to tensile forces. Consider the part of the bar shown in Figure 173 (*b*). For the bar to be in equilibrium the sum of the internal forces must equal the applied force.

That is,
sum of the internal forces = applied force
or,
force per unit area × resisting cross-sectional area = applied force.

But, the force per unit area is known as stress; therefore the equation can be written as
stress × resisting cross-sectional area = applied force

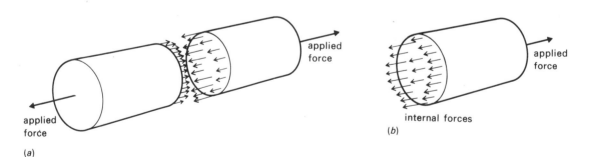

(a)

(b)

Figure 173   *Internal forces*

Rearranging this equation, the intensity of stress can be calculated from the equation

$$\text{stress} = \frac{\text{applied force}}{\text{resisting cross-sectional area}}$$

The symbol used for normal stress is the Greek letter σ (sigma), so that the equation may be written as:

$$\sigma = \frac{F}{A} \quad \text{where} \quad F = \text{applied force,}$$
$$A = \text{resisting cross-sectional area.}$$

The basic unit for stress is the N/m^2 but the most commonly used units are N/mm^2 or MN/m^2.

In all the examples considered so far the applied forces have been shown acting in pairs, see Figures 159 to 163. In this topic area consideration will be given only to bodies which are in equilibrium. For a body to be in equilibrium the magnitude of each of the two applied forces must be the same. Consider two teams pulling against each other in a tug of war competition. If the two teams apply the same total force to each end of the rope then the rope is in a state of equilibrium. When calculating the value of the intensity of stress in a body which is in a state of equilibrium under the action of a pair of applied forces, only one value of the applied force is used.

Consider the following example

Example 1

A mild steel bolt 40 mm in diameter has to carry tensile forces of 60 kN. Calculate the tensile stress in the bolt because of these forces.

Figure 174 (a) shows a sketch of the bolt. To calculate stress use the equation:

$$\text{stress} = \frac{\text{applied force}}{\text{resisting cross-sectional area}}$$

Note that, in calculating the value of the stress only one of the 60 kN forces is used (see Figure 174 (b)). The resisting area is the area normal (i.e. at 90°) to the axis of the applied force.

$$\text{resisting area} = \frac{\pi d^2}{4} = \frac{\pi \times 40^2}{4} \ [mm^2]$$
$$= 1256.6 \ mm^2$$

Therefore the stress in the bolt is:

$$\sigma = \frac{F}{A}$$

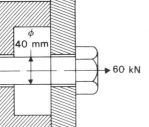

internal force = stress × resisting cross-sectional area

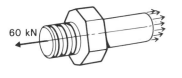

Figure 174 *Bolt in tension*

$$\sigma = \frac{60 \times 10^3}{1256.6} \frac{[N]}{[mm^2]}$$

$$\sigma = 47.7 \, N/mm^2 \quad or \quad 47.7 \, MN/m^2$$

Tensile forces produce tensile stress in a material, and compressive forces produce compressive stress in a material. Because these forces are normal forces, the associated stresses are known as normal stresses. When calculating normal stresses, the area used in the calculations is the area normal (i.e. at $90°$) to the line of action of the forces.

Shearing forces produce shearing stress in a material. When calculating shearing stress the area used is the area parallel to the line of action of the force. As was mentioned earlier, a shearing force is also known as a transverse force; similarly shear stress is also known as transverse stress. The symbol used for transverse stress is the Greek letter τ (tau). Figure 175 represents a piece of material which has been sheared. The area which has been sheared is shown shaded. This is the resisting cross-sectional area and is used to calculate the value of the shear or transverse stress.

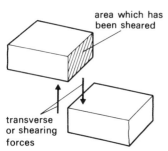

Figure 175   *Shear failure*

Example 2

The support leg of a shaping machine is 100 mm diameter. During a certain machining operation, the compressive forces on the leg are 8 kN. Calculate the compressive stress in the leg due to these forces.

Figure 176 shows a sketch of the leg. To calculate the stress use the equation:

$$stress = \frac{applied \ force}{resisting \ cross\text{-}sectional \ area}$$

The resisting area = $\dfrac{\pi d^2}{4} = \dfrac{\pi \times 100^2}{4} \, mm^2]$

$$= 7854 \, mm^2$$

Therefore the compressive stress in the leg is

$$\sigma = \frac{F}{A}$$

$$\sigma = \frac{8 \times 10^3}{7854} \frac{[N]}{[mm^2]}$$

$$\sigma = 1.02 \, N/mm^2 \quad or \quad 1.02 \, MN/m^2$$

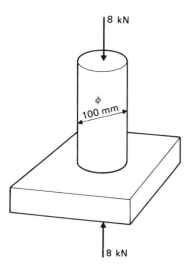

Figure 176   *Machine leg in compression*

Example 3

A piece of metal strip 50 mm wide by 3 mm thick is to be sheared on a guillotine shear. If the stress required to shear the metal is 300 MN/m², calculate the forces required at the blade of the shear.

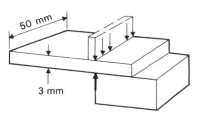

Figure 177   *Plate in shear*

Figure 177 illustrates the problem. The relationship between stress, applied force and resisting area is given by the equation:

$$\text{stress} = \frac{\text{applied force}}{\text{resisting cross-sectional area}}$$

Rearranging the equation to give the applied force:

applied force = stress × resisting cross-sectional area

In this case the area to be sheared = 50 × 3 = 150 mm²
Therefore the applied force is:

$$\begin{aligned}
\text{force} &= \text{stress} \times \text{area} \\
&= 300 \left[\frac{MN}{m^2}\right] \times 150 \ [mm^2] \ , \\
&= 300 \times 10^6 \left[\frac{N}{m^2}\right] \times \frac{150}{10^6} [m^2] \ , \\
&= 300 \times 150 \ N \ , \\
\text{force} &= 45 \ kN
\end{aligned}$$

Example 4

In a particular punching operation 20 mm diameter holes are being punched in a sheet of brass 3 mm thick. If the shear strength of the brass is such that a stress of 180 MN/m² is required to punch a hole in the brass, calculate:

(a)   the smallest force that has to be applied to the punch in order to cut the metal,

(b)   the compressive stress in the punch shank if the shank is 18 mm diameter.

Figure 178 illustrates the problem.

(a)   Area to be cut by the punch   =   circumference of hole × thickness of sheet
$$\begin{aligned}
&= \pi \times D \times \text{thickness} \\
&= \pi \times 20 \times 3 \ [mm] \times [mm] \\
&= 188.5 \ mm^2
\end{aligned}$$

Applied force = stress × resisting cross-sectional area

$$= 180 \left[\frac{MN}{m^2}\right] \times 188.5 \ [mm^2]$$

changing the units for the area:

$$F = 180 \left[\frac{MN}{m^2}\right] \times \frac{188.5}{10^6} [m^2]$$

$$F = 0.034 \ MN = 34 \ kN$$

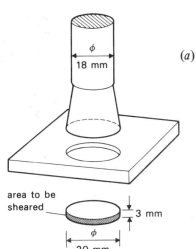

Figure 178   *Example 4*

(b)   Compressive stress in punch shank $= \dfrac{F}{A}$

resisting area $= \dfrac{\pi D^2}{4} = \dfrac{\pi \times 18^2}{4} = 254.5 \text{ mm}^2$

compressive stress $= \dfrac{F}{A}$

$= \dfrac{34 \quad [\text{kN}]}{254.5 \; [\text{mm}^2]}$

$= \dfrac{34 \times 10^3 \quad [\text{N}]}{254.5 \times 10^{-6} \quad [\text{m}^2]}$

$= 133.6 \times 10^6 \text{ N/m}^2$

$= 133.6 \text{ MN/m}^2$

Example 5

Figure 179 shows a rivet clamping two plates together. The rivet has to withstand maximum forces of 10 kN, and the stress set up in it must not exceed 45 MN/m². Calculate the smallest diameter rivet that may be used. What type of stress is caused in the rivet?

The rivet is subjected to a shear stress.

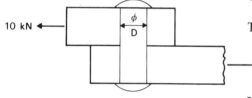

Figure 179   *Example 5*

The formula: stress $= \dfrac{\text{applied force}}{\text{resisting cross-sectional area}}$

or $\tau = \dfrac{F}{A}$

rearranged gives:

resisting cross-sectional area $= \dfrac{\text{applied force}}{\text{stress}}$

or $A = \dfrac{F}{\tau}$

where $F = 10$ kN and $\tau = 45$ MN/m² (N/mm²)

therefore, $A = \dfrac{10 \quad [\text{kN}]}{45 \; [\text{N/mm}^2]}$

$A = \dfrac{10 \times 10^3}{45} \; [\dfrac{\text{N}}{\text{N}}\text{mm}^2]$

$A = 222.2 \text{ mm}^2$

For the rivet, the resisting cross-sectional area is given by the equation:

$$A = \dfrac{\pi D^2}{4}$$

Therefore, $222.2 = \dfrac{\pi D^2}{4}$

hence $D^2 = \dfrac{4 \times 222.2}{\pi} = 282.9 \text{ mm}^2$

therefore $D = \sqrt{282.9} = 16.82 \text{ mm}$

The minimum rivet diameter required is 16.82 mm; the next largest stocked size would be used.

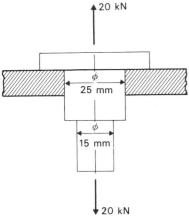

Figure 180   *Example 6*

Example 6

Figure 180 shows part of a lifting device. Calculate the stress in each portion of the pin if the applied forces are 20 kN.

Use the formula $\sigma = \dfrac{F}{A}$

For the 15 mm diameter portion, $A = \dfrac{\pi \times 15^2}{4} = 176.7 \text{ mm}^2$

Therefore the stress is: $\sigma = \dfrac{20 \times 10^3}{176.7} \dfrac{[\text{N}]}{[\text{mm}^2]}$

$= 113.2 \dfrac{[\text{N}]}{[\text{mm}^2]}$

For the 25 mm diameter portion, $A = \dfrac{\pi \times 25^2}{4} = 490.9 \text{ mm}^2$

Hence the stress is: $\sigma = \dfrac{20 \times 10^3}{490.9} \text{ N/mm}^2$

$\sigma = 40.74 \text{ N/mm}^2$

Note that the force is the same in both portions of the pin, but the stress is different. The larger cross-sectional area is subjected to a smaller stress.

Strain

It has already been stated that a solid distorts enough to build up forces which just counter the external forces applied to it. The distortion caused is a measure of the strain brought about by the application of the external forces. Strain is the change in dimension per unit original dimension and is calculated using the equation:

$$\text{strain} = \frac{\text{change in length}}{\text{original length}}$$

Tensile or compressive strain is denoted by the symbol ϵ (epsilon).

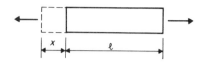

(a) tensile strain

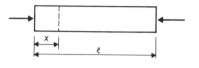

(b) compressive strain

Figure 181   *Tensile and compressive strain*

From Figure 181 (a) and (b), the strain due to tensile or compressive forces is given by:

$$\epsilon = \frac{x}{\ell}$$ where ϵ = tensile or compressive strain
x = change in length
ℓ = original length

Strain is a ratio of lengths, and hence has no units. On occasions, it is expressed as a percentage of the original length.

Example 7
Calculate the compressive strain if a component 150 mm long is compressed by 0.05 mm under the action of compressive forces.

$$\text{compressive strain} = \frac{\text{change in length}}{\text{original length}}$$

$$\text{or } \epsilon = \frac{x}{\ell}$$

$$\epsilon = \frac{0.05 \text{ mm}}{150 \text{ mm}}$$

$$\epsilon = 0.000\,33$$

Expressed as a percentage, the % strain $= \dfrac{0.05}{150} \times 100 = 0.033\%$

Example 8
If a bar 3 m long is stretched until its length has increased by 0.9 mm, calculate the tensile strain and express this as a percentage strain value.

$$\text{tensile strain} = \frac{\text{change in length}}{\text{original length}}$$

$$\text{i.e.} \quad \epsilon = \frac{x}{\ell}$$

$$\text{or} \quad \epsilon = \frac{0.9}{3} \frac{[\text{mm}]}{[\text{m}]} = 0.000\,3$$

As a percentage value, the % strain = 0.03%

Example 9
Calculate the extension of a 2 m length of bar if the percentage strain is 0.05%.

If the percentage strain = 0.05%

then the strain $\quad \epsilon = \dfrac{0.05}{100} = 0.000\,5$

Now, strain $= \dfrac{\text{change in length}}{\text{original length}}$

which, rearranged, gives:

change in length $=$ strain $\times$ original length
$$= 0.000\,5 \times 2\,000 \ [\text{mm}]$$
$$= 1 \text{ mm}$$

Self-assessment questions

3   Read each of the following statements carefully and then complete each statement by adding the word/words which is/are applicable.
    (i)   In order to produce the necessary internal force to resist applied forces, a solid must _____ .
    (ii)   Stress is defined as the _____ .
    (iii)   The equation relating stress, applied force and resisting cross-sectional area is stress = _____ .
    (iv)   The basic S I unit used for stress is _____ .
    (v)   Strain has no units and may be expressed as a _____ .
    (vi)   The equation relating strain, change in dimension and original dimension is given by strain = _____ .

4   Figure 182 shows four bars of steel, A, B, C and D. The four bars are made from identical material and are all subjected to the same applied forces F. All four bars have a square cross-section. List the four bars in order of the value of the intensity of stress, starting with the bar which will have the greatest value of stress.

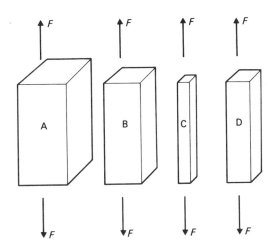

Figure 182   *Self-assessment question 4*

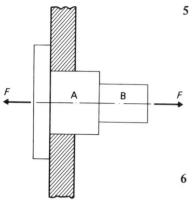

Figure 183   *Self-assessment question 5*

5   Figure 183 shows a stepped pin. The cross-sectional area of the small diameter portion is exactly half that of the large diameter portion. If the stepped pin is subjected to applied tensile forces F as shown, which of the following six statements are true?
(i)   stress in portion A = stress in portion B,
(ii)   stress in portion A = 2 × stress in portion B,
(iii) stress in portion A = ½ × stress in portion B,
(iv)   force in portion A = force in portion B,
(v)   force in portion A = 2 × force in portion B,
(vi)   force in portion A = ½ × force in portion B.

6   A 10 mm diameter punch has a punch force of 15 kN acting on it. Calculate the stress in the punch, and state the type of stress to which it is subjected.

7   A bar of steel 30 mm diameter is to be turned between centres. The forces applied to the ends of the bar when the tailstock is tightened is 300 N. State the kind of stress which results in the bar, and calculate its magnitude.

8   A 25 mm diameter bolt is tightened such that it is subjected to a stress of 15 MN/m². What are the forces required to cause this stress?

9   Calculate the strain in each of the following cases, and express the strain as a percentage value.
(*a*) A 2000 mm bar is extended by 1 mm,
(*b*) A 4 m bar is compressed by 1.6 mm.

10   Calculate the change in length of a bar of material 2 m long if it has:
(*a*) 0.05% strain,
(*b*) 0.1% strain.

11   When a hacksaw blade is tightened in its frame the tensile stress set up in it must not exceed 15 MN/m². The blade is 12 mm wide by 0.7 mm

---

Solutions to self-assessment questions

3   (i)   distort, i.e. change shape.
     (ii)   force per unit cross-sectional area.

     (iii) stress = $\dfrac{\text{applied force}}{\text{resisting cross-sectional area}}$

     (iv)   N/m² is the basic unit, but N/mm² or MN/m² are more common.
     (v)   fraction, or ratio, or percentage value.

     (vi)   strain = $\dfrac{\text{change in length}}{\text{original length}}$

4   The order is C, D, B, A. Remember the smaller the area the greater the stress, for a given applied force.

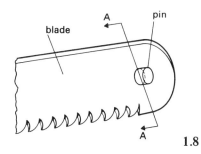

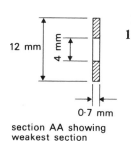

12 mm

4 mm

0·7 mm

section AA showing
weakest section

Figure 184   *Tensile stress in hacksaw blade*

thick, and the pin holes are 4 mm diameter. Figure 184 illustrates part of a hacksaw blade. Calculate the greatest permissible tightening forces that can be applied to the blade. Calculate the stress in the pin, assuming it to be 4 mm diameter. To what kind of stress is the pin subjected?

After reading the following material, the reader shall:

1.8   Draw graphs of force against extension and stress against strain for an elastic material.

1.9   Define Young's modulus and relate it to the 'stiffness' of a material.

1.10   Solve problems involving stress, strain and Young's modulus.

Elasticity is defined as the ability of a material to return to its original dimensions when applied forces are removed. Elastic behavior is shown by the majority of engineering materials.

As has been mentioned earlier, for a material to be able to resist applied forces it must be distorted. The most common method of representing the load carrying characteristics of an engineering material is to draw the force-extension or stress-strain curve. Later in this topic area the procedure for carrying out a tensile test to destruction will be studied in detail. In order to illustrate several important points in the study of the effect of forces on materials, consider the table of results shown below. The table contains values of gradually applied axial tensile forces and the related extensions for a specimen of hardened brass.

axial force (kN)	0	2	4	6	8	10	12	14	16
extension (mm)	0	.0032	.0064	.0096	.0128	.0159	.019	.022	.0251

axial force (kN)		20	21	23	24	26	28	30
extension (mm)		.0335	.0378	.0505	.06	.083	.106	.145

Figure 185 shows the force-extension curve for the brass material. The straight line portion of the curve represents a very important relationship which was first identified by the physicist, architect and engineer Robert Hooke (1635-1702). The relationship is known as *Hooke's law* and is usually stated as: Extension is directly proportional to the force producing it provided that the material remains within the elastic limit.

It will be remembered that elasticity is defined as the ability of a material to return to its original dimensions. If the external forces acting on a body are so large that the body does not return to its original dimensions after the forces are removed, then the elastic limit for the

Solutions to self-assessment questions

5 ' The only two true statements are that:

(iii) stress in portion A = ½ × stress in portion B,

and (iv) force in portion A = force in portion B.

The stress in B is twice that in A because the area of B is only half that of A. In a stepped bar the force in each section does not change, only the stress varies.

6   Using $\sigma = \dfrac{F}{A}$, then $\sigma = \dfrac{15 \times 10^3 \quad [N]}{(\pi \times 10^2)/4 \quad [mm^2]}$

Therefore the compressive stress = 191 N/mm²   = 191 MN/m²

7   Using $\sigma = \dfrac{F}{A}$, then $\sigma = \dfrac{300 \quad [N]}{\pi \, 30^2/4 \quad [mm^2]}$

Therefore the compressive stress = 0.424 N/mm²   = MN/m²

8   Rearrange equation $\sigma = \dfrac{F}{A}$ to give $F = \sigma \times A$

forces in the bolt $= 15[\dfrac{N}{mm^2}] \times \pi \dfrac{25^2}{4} \ [mm^2]$

$= 7\,363 \ N = 7.363 \ kN$

9   (a) strain $= \dfrac{\text{change in length}}{\text{original length}}$

$\epsilon = \dfrac{1}{2\,000} = 0.000\,5$

As a percentage value, % strain = 0.000 5 × 100 = 0.05%

(b) compressive strain $= \dfrac{1.6}{4\,000} = 0.000\,4$

Therefore % strain = 0.000 4 × 100 = 0.04%

10 (a) Rearranging the equation: strain $= \dfrac{\text{change in length}}{\text{original length}}$

to give: change in length = strain × original length

Then, change in length $= \dfrac{0.05}{100} \times 2\,000 = 1 \ mm$

(b) Again using change in length = strain × original length

change in length $= \dfrac{0.1}{100} \times 2\,000 = 2 \ mm$

11 Use the equation $F = \sigma \times A$ where A is the area at the weakest part of the blade.

Now $A = (8 \times 0.7) \ mm^2 = 5.6 \ mm^2$

Hence the tightening forces $= \sigma \times A$

$= 15 \ [\dfrac{N}{mm^2}] \times 5.6 \ [mm^2]$

$= 84 \ N$

The pin is subjected to shear stress.

Using $\tau = \dfrac{F}{A} = \dfrac{84 \quad [N]}{\pi \, 4^2/4 \quad [mm^2]}$

$= 6.68 \ N/mm^2$

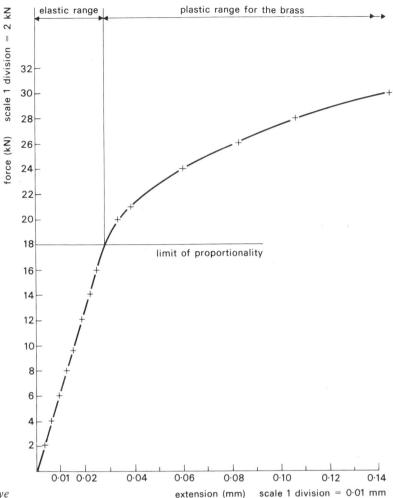

Figure 185   *Force-extension curve*

material has been exceeded. The elastic limit force for a material is the maximum force which can be exerted on a body without causing a permanent distortion. On the force-extension curve, the range of forces and extensions for which the material can be considered to be elastic corresponds to the straight line portion of the curve. Beyond the elastic limit the material enters the plastic stage. The plastic range for a metal corresponds very closely to the curved portion of the force-extension curve. If a material is subjected to applied forces which are within the plastic range for the material then the material suffers a permanent distortion.

The point on the force-extension curve at which the straight line relationship ends is strictly speaking the *limit of proportionality. The limit* of proportionality may differ slightly from the elastic limit of the material. For a material such as steel, however, the two limits are practically the same.

For the force-extension curve shown in Figure 185 the limit of proportionality for the brass is 18 kN. The straight line relationship described by Hooke's law is directly applicable to most metals, but it does not apply to all materials. Many polymers do not conform to Hooke's law.

Engineers and designers need to know the maximum force which may be applied in service to a particular component. This force can then be compared with the force corresponding to the elastic limit of the material.

Summarizing
Elasticity is defined as the ability of a material to return to its original dimensions when the applied forces are removed. The *elastic region* of a force-extension curve for a metal corresponds very closely to the straight-line portion of the graph. The straight line relationship between the applied force and the extension ends at *the limit of proportionality*. The *plastic region* of a force-extension curve for a metal corresponds very closely to the curved portion of the graph.

Stress-strain relationship
Consider the table of results shown below. These results, already used to obtain the force-extension curve shown in Figure 185, are obtained from a hardened brass specimen which has a cross-sectional area of 320 mm². The original length of the specimen is 50 mm.

axial force (kN)	0	2	4	6	8	10	12	14
stress = $\frac{F}{A}$ (N/mm²)	0	6.25	12.5	18.75	25	31.25	37.5	42.5
extension (mm)	0	0.0032	0.0064	0.0096	0.0128	0.0159	0.019	0.022
strain = $x/\ell \times 10^{-4}$	0	0.64	1.28	1.92	2.56	3.18	3.8	4.4

axial force (kN)	16	20	21	23	24	26	28	30
stress = $\frac{F}{A}$ (N/mm²)	50	62.5	65.6	71.9	75	83.75	87.5	93.75
extension (mm)	0.0251	0.0335	0.0378	0.0505	0.06	0.083	0.106	0.145
strain = $x/\ell \times 10^{-4}$	5.2	6.7	7.56	10.1	12	16.6	21.2	29

Figure 186 shows the stress-strain curve for the brass specimen. The shape of the force-extension and the stress-strain curves for any specimen is the same, because all that has been done is to change the force values into stress values, by dividing by a constant (the cross-sectional area of the specimen), and to change the extension values into

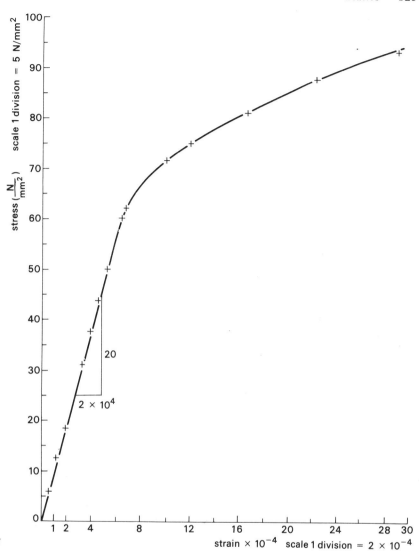

Figure 186   *Stress-strain curve*

strain values by dividing by the original length of the specimen. As with the force-extension curve, the straight line portion of the stress-strain curve may be said to represent the elastic range of the material. The stress ceases to be directly proportional to strain when the stress value reaches 60 N/mm² and the strain value is 6.4×10^{-4}. Thus the limit of proportionality is 60 N/mm².

The value of the slope of the straight line portion of the stress-strain curve for a material is a fundamental property of the material itself. The value of this constant is called Young's modulus of elasticity after Thomas Young (1773-1829) who first realised its importance. The con-

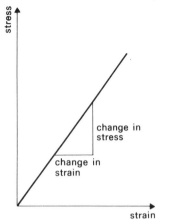

Figure 187   *Slope of straight line portion of stress-strain curve*

stant is sometimes called simply 'Young's modulus', or the 'modulus of elasticity', and is represented by the symbol E.

Figure 187 shows part of the straight line portion of a stress-strain curve for a material.

Young's modulus of elasticity = slope of straight line portion of graph

$$= \frac{\text{change in stress}}{\text{change in strain}}$$

$$\text{or } E = \frac{\text{change in stress}}{\text{change in strain}}$$

Since the strain is a number, the units in which Young's modulus of elasticity is measured are the same as those for stress; i.e. force per unit area. *Remember this relationship is true only for the straight line portion of the graph, i.e. up to the limit of proportionality.*

Consider the two stress-strain curves for metals A and B shown in Figure 188. For a stress value of, say, 40 N/mm², a value which is below the elastic limit of both metals, the resulting strains are 0.0004 and 0.0008 in metals A and B respectively. The material which exhibits the smaller value of strain is said to be stiffer than the other material. Stiffness may be defined as the ability of a material to resist deformation. For example, steel is stiffer than rubber.

Which of the two metals shown in Figure 188 has the larger value of Young's modulus? The answer is metal A because it has the bigger value for the slope of the straight line portion of the stress-strain curve. Young's modulus is one measure of the stiffness of a material.

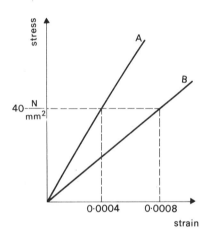

Figure 188   *Comparison of stress-strain curves*

Examples of the values of Young's modulus for a range of different materials are shown in the table opposite.

material	approximate value for E (GN/m^2)
rubber	7×10^{-3}
steel	210
diamond	1 200
wood	14
aluminium	72
unreinforced plastic	1.4

Referring to Figure 186, the value of Young's modulus of elasticity can be calculated using values taken from the graph. By constructing a right angled triangle as shown and reading off the values of change of stress and change of strain, the value for Young's modulus may be calculated.

From the graph, change of stress = 20 N/mm² and change of strain = 2×10^{-4}.

Therefore, $E = \dfrac{20}{2 \times 10^{-4}}$ [N/mm^2]

$= 100$ kN/mm^2 or 100 GN/m^2

Consider some examples where the value of Young's modulus can be used.

Example 10

A component in a machine linkage is 20 mm diameter and 400 mm long. The component is made from steel which has a value of Young's modulus of 210 GN/m^2 and has to withstand a maximum tensile stress of 14 N/mm^2 which is below the limit of proportionality. Calculate the maximum change in length of the 400 mm long component.

To find the change in strain use the relationship

$$E = \frac{\text{change in stress}}{\text{change in strain}}$$

which rearranges to give

$$\text{change in strain} = \frac{\text{change in stress}}{E}$$

The change in stress is from 0 to 14 N/mm^2, i.e. 14 N/mm^2; therefore the change in strain

$$= \frac{14 \quad [\text{N/mm}^2]}{210 \times 10^3 \quad [\text{N/mm}^2]}$$

$$= 6.667 \times 10^{-5}$$

To find the extension of the component use the relationship

$$\text{strain} = \frac{\text{change in length}}{\text{original length}}$$

which rearranges to give

change in length $=$ strain $\times$ original length
$= 6.667 \times 10^{-5} \times 400$ mm
$= 0.0267$ mm

The maximum change in length occurs when the component is subjected to a stress of 14 N/mm^2. At this stress the length of the component is increased by 0.0267 mm.

Example 11

A hydraulic press has four vertical tie bars 100 mm diameter and 2.5 m long fixed to a bedplate and supporting cylinders as shown in Figure 189. The four tie bars are subjected to stresses which vary from zero to a maximum of 20 MN/m^2 which is below the limit of proportionality.

tie bars

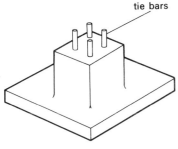

Figure 189   *Example 11*

The tie bars are made from steel having a value of Young's modulus of 200 GN/m^2. Calculate the maximum forces that are applied by the press and the corresponding change in length of the tie bars.

To find the forces that may be applied by the press use the relationship

$$\text{stress} = \frac{\text{applied force}}{\text{resisting area}}$$

which rearranges to give

applied force = stress $\times$ resisting area

The resisting area is the total cross-sectional area of the four tie bars:

$$\text{resisting area} = 4 \times \pi \frac{d^2}{4}$$

$$= 4 \times \pi \frac{100^2}{4} \text{ mm}^2$$

$$= 31\,416 \text{ mm}^2$$

Therefore, the applied forces $= 20 [\frac{N}{mm^2}] \times 31\,416 [mm^2]$

$$= 628.32 \text{ kN}$$

The maximum forces which may be applied by the press are 628.32 kN. To find the change in strain use the relationship

$$E = \frac{\text{change in stress}}{\text{change in strain}}$$

which rearranges to give

$$\text{change in strain} = \frac{\text{change in stress}}{E}$$

$$= \frac{20 \times 10^6 \quad [\text{N/m}^2]}{200 \times 10^9 \quad [\text{N/m}^2]}$$

$$= 1 \times 10^{-4}$$

To find the extension of the tie bars use the relationship

$$\text{strain} = \frac{\text{change in length}}{\text{original length}}$$

which rearranges to give

change in length = strain $\times$ original length
$$= 1 \times 10^{-4} \times 2.5 \text{ [m]}$$
$$= 1 \times 10^{-4} \times 2.5 \times 10^3 \text{ [mm]}$$
$$= 0.25 \text{ mm}$$

When the tie bars are subjected to a stress of 20 MN/m^2 they extend by 0.25 mm.

Self-assessment questions

Read each of the following statements carefully and complete the statements by adding the missing word/words.

12   (i)   If an elastic material is subjected to applied forces which are below the elastic limit of the material, when the forces are removed the material returns _____ .

(ii)   On the force-extension, or the stress-strain, graph for a metal, the elastic region corresponds very closely to the _____ portion of the graph.

(iii) On the force-extension, or the stress-strain, graph for most metals, the plastic region corresponds very closely to the _____ portion of the graph.

(iv)   The straight line relationship between applied force and extension ends at the _____ .

(v)   Young's modulus of elasticity is a measure of the _____ of a material.

(vi)   Young's modulus of elasticity is defined as the slope of _____ _____ .

(vii) The units for E are the same as those used for _____ .

(viii) The value of E for a metal is calculated by using values from the _____ portion of the stress-strain curve.

13   Given that the value of E for steel is 210 GN/m^2, and that the value of E for aluminium is 70 GN/m^2, what is the ratio of the stiffness of steel to the stiffness of aluminium?

14   Study the following list of materials and their values of elasticity, and list the materials in order of stiffness, starting with the stiffest material first.

code	material	modulus of elasticity (GN/m^2)
A	wrought iron	190
B	steel	210
C	cast iron	110
D	brass	83
E	copper	96
F	wood	10

15   From the straight line portion of a force-extension graph for a given material it is found that a change in force of 25 kN produces a change in extension of 0.106 mm. If the cross-sectional area of the specimen is 100 mm^2 and the original length of the specimen is 50 mm, calculate the value of the modulus of elasticity for the material.

16   The following force and extension values are obtained when a steel bar is subjected to a series of tensile forces.

force (kN)	0	5	10	15	20	25	30
extension (mm)	0	.0118	.0236	.0355	.0472	.059	.071

The cross-sectional area of the steel bar is 100 mm², and the extensions are based on an original length of 50 mm. Plot a stress-strain graph for this material and hence determine the modulus of elasticity.

17   The following force and extension values are obtained when an aluminium alloy specimen of 11.28 mm diameter and length 50 mm is subjected to a series of tensile forces.

force (kN)	0	5	10	15	20	25	30
extension (mm)	0	.0354	.07	.1065	.1416	.177	.213

Plot a stress-strain graph for this material, and hence determine the modulus of elasticity for the aluminium alloy.

18   A mild steel bush 50 mm outside diameter, 25 mm bore, and 200 mm long is placed vertically in a press and subjected to vertical forces of 50 kN. Calculate the stress in the material and the decrease in the

---

Solutions to self-assessment questions

12   (i)   to its original dimensions.
(ii)   straight line.
(iii)   curved.
(iv)   limit of proportionality.
(v)   stiffness.
(vi)   straight line portion of the stress-strain graph.
(vii)   stress.
(viii) straight line.

13 $\dfrac{\text{stiffness of steel}}{\text{stiffness of aluminium}} = \dfrac{E_S}{E_A} = \dfrac{210}{70} = \dfrac{3}{1}$, i.e. 3:1 .

14   Order of stiffness. B, A, C, E, D, F. Remember that the value of the modulus of elasticity is a measure of the stiffness of a material. The higher the value for E, the stiffer the material.

15 $E = \dfrac{\text{change in stress}}{\text{change in strain}}$, where stress $= \dfrac{\text{force}}{\text{area}}$

and strain $= \dfrac{\text{extension}}{\text{original length}}$

Therefore change in stress $= \dfrac{25 \times 10^3}{100} \dfrac{[\text{N}]}{[\text{mm}^2]} = 250 \text{ N/mm}^2$,

and change in strain $= \dfrac{0.106}{50} \dfrac{[\text{mm}]}{[\text{mm}]} = 2.12 \times 10^{-3}$

Therefore $E = \dfrac{250}{2.12 \times 10^{-3}}$ N/mm² $= 117.9$ kN/mm² or 117.9 GN/m² length of the bush, assuming that the stress is below the limit of proportionality stress. The value of the modulus of elasticity for the bush material is 210 GN/m². Assume that steel in compression behaves in a manner similar to steel in tension.

After reading the following material, the reader shall:

1.11 Describe a tensile test to destruction.

1.12 Describe the form of stress-strain graphs for brittle and ductile materials.

1.13 Define the terms: ductility, brittleness, hardness, limit of proportionality, elastic limit, strength.

1.14 Calculate the following values for a given material:
(i) modulus of elasticity,
(ii) stress at limit of proportionality,
(iii) ultimate stress,
(iv) percentage elongation,
(v) percentage reduction in area.

1.15 Interpret stress-strain curves to predict the properties of a material.

Several references have been made to force-extension and stress-strain graphs for materials. It will be appreciated that both of these graphs present the designer or engineer with a picture of how a material behaves under the action of applied tensile forces. To enable comparisons to be made between the behaviour of different materials it is beneficial to standardise the method of testing. In fact two of the earliest British Standards, BSS 3 and BSS 18 dealt with materials testing. Machines capable of carrying out tensile and compressive tests on materials have been in use for over 100 years. The tensile test in particular has been widely used by industry to determine the properties of materials. In addition to the value of the modulus of elasticity there are other properties of a material which can be determined from a tensile test. These other properties (which will be discussed more fully later in this topic area) include the strength and ductility of a material. Variations in the composition and the heat treatment a material has received can cause considerable variations in these properties.

In order that the results of tests may be used for the purposes of comparison, standard shapes and sizes are required for the samples used as test pieces. Figure 190 shows the main dimensions of one very commonly used standard tensile test specimen.

A diameter of 11.283 mm is chosen in order that the cross-sectional area of the test piece is 100 mm². Other diameters may be used for other standard test pieces.

To perform a tensile test to destruction, a tensile testing machine is

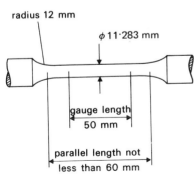

radius 12 mm

φ 11·283 mm

gauge length
50 mm

parallel length not
less than 60 mm

Figure 190   *A standard tensile test piece*

required. Full details of those in current production are to be found in sales literature.

A tensile testing machine, in its essentials, consists of machinery for exerting a pull on a test piece coupled to a device which measures the force. The straining mechanism may consist of a screw, driven in an axial direction by the rotation of a nut or the mechanism may be hydraulic, consisting of a cylinder with a piston carrying one of the test piece grips.

Most modern testing machines are built to perform both compressive and tensile tests, and such a machine is termed universal, to distinguish it from machines of more restricted scope.

In addition to the application and measurement of the applied forces, an extension measuring instrument is required. Many different kinds of extensometer are available. They range from the Lamb's roller extensometer, which is an optical extensometer and requires a great deal of

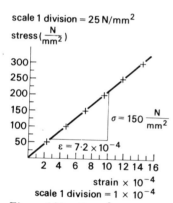

scale 1 division = 25 N/mm^2

Figure 191   *Solution to self-assessment question 16*

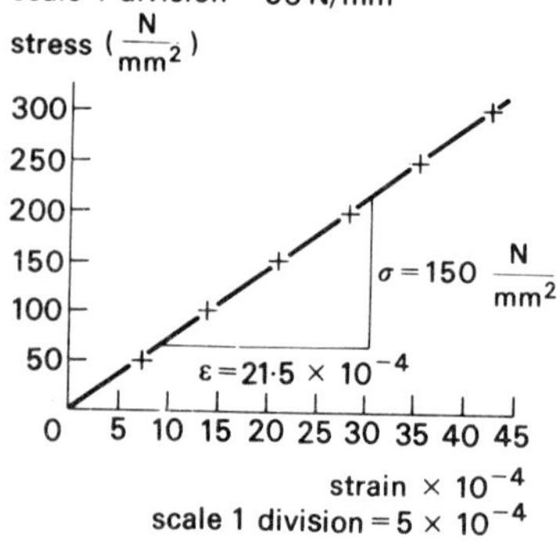

scale 1 division = 50 N/mm^2

Figure 192   *Solution to self-assessment question 17*

Solutions to self-assessment questions

16   The graph of stress against strain is shown in Figure 191. Using this graph to calculate the value of the modulus of elasticity:

from the graph, change is stress = 150 N/mm^2 and change in strain = 7.2×10^{-4}

Therefore $E = \dfrac{150}{7.2 \times 10^{-4}}$ N/mm^2

$E = 208$ kN/mm^2 or 208 GN/m^2

17   The graph of stress-strain is shown in Figure 192. Using the values from the graph:

$E = \dfrac{150}{21.5 \times 10^{-4}}$ N/mm^2

$= 69.77$ kN/mm^2 or 69.77 GN/m^2

18   Stress, $\sigma = \dfrac{F}{A}$, where $F = 50$ kN

and $A = \dfrac{\pi}{4}(50^2 - 25^2) = 1472.6$ mm^2

Therefore $\sigma = \dfrac{50 \times 10^3}{1472.6}$ N/mm$^2 = 33.95$ N/mm^2

Change in strain $= \dfrac{\text{change in stress}}{E} = \dfrac{33.95 \text{ N/mm}^2}{210 \times 10^3 \text{ N/mm}^2}$

$= 0.1617 \times 10^{-3}$

To find the amount of compression, use

$\epsilon = \dfrac{x}{\ell}$ rearranged to give

$x = \epsilon \times \ell$

$= 0.1617 \times 10^{-3} \times 200$ [mm]

$= 0.0323$ mm

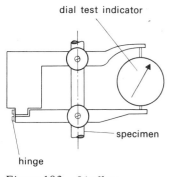

dial test indicator

specimen

hinge

Figure 193 *Lindley extensometer*

care in setting up, to the Lindley extensometer shown in Figure 193, which is of a more robust construction. Much of the laborious plotting of force-extension or stress-strain graphs can be avoided by the use of attachments to testing machines which automatically draw the graph as the force is applied.

Irrespective of the type of testing machine and the type of extensometer used, the procedure for performing a tensile test to destruction can be generalized as follows:

1 Measure the initial diameter and gauge length of the specimen and record the values.

2 Insert the specimen in the testing machine and attach the extensometer. Check both the force dial recorder and the extensometer for initial or zero readings. If either does not record zero they may be adjusted to give a zero reading.

3 Choosing a suitable range of force which will depend upon the type of material being tested and on the machine being used, increase the force in suitable increments, noting the force and extension value at each increment. Again, depending upon the type of extensometer being used, the extensometer may have to be removed before the specimen breaks. If the extensometer has to be removed, the extensions during the final part of the test may be measured by the use of vernier calipers. If in doubt, the extensometer should be removed when the material reaches its limit of proportionality. The force is increased until the specimen breaks. Care should be taken to determine the actual breaking force.

4 Remove the specimen from the testing machine and bring the two separate pieces together to form a complete specimen again, measure (i) the final extension, and (ii) the diameter of the specimen at the smallest section.

These measurements are illustrated in Figure 194. Having obtained the data detailed above, various calculations can be made, as illustrated in the following example.

diameter at smallest cross-section

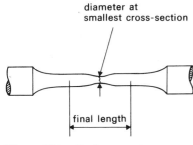

final length

Figure 194 *Broken specimen*

Example 12

In a tensile test to destruction on an aluminium alloy specimen, 11.283 mm diameter, guage length 50 mm, the following extensions were obtained for the given forces:

force (kN)	0	2	4	6	8	10	12	14
extension (mm)	0	.0133	.0266	.04	.0532	.067	.08	.093
force (kN)	16	18	19	20	24	28	29	28
extension (mm)	.1064	.15	.234	.5	1.3	3.4	4.4	8.4

The other results noted were as follows:

The maximum force recorded was 29.6 kN.

The value of the force when fracture occurred was 24 kN.

The total extension was 10.5 mm.

The final diameter was 9.17 mm.

Using the above results, draw a stress-strain graph for the specimen. Using the graph and the results determine:

(i)   the modulus of elasticity for the material,
(ii) the limit of proportionality stress,
(iii) the ultimate stress,
(iv) the percentage elongation,
(v)   the percentage reduction in area.

To obtain the stress and strain values for the test, use the equations:

$$\text{stress} = \frac{\text{applied force}}{\text{resisting cross-sectional area}}$$

$$\text{and, strain} = \frac{\text{extension}}{\text{original length}}$$

For the specimen under test:

$$\text{the resisting cross-sectional area} = \frac{\pi D^2}{4} = \frac{\pi \times 11.283^2}{4}$$

$$= 100 \text{ mm}^2$$

and the original length = gauge length = 50 mm

The table of stress-strain values will therefore be:

stress $= \dfrac{\text{force} \times 10^3}{100}$ N/mm^2	0	20	40	60
strain $= \dfrac{x}{50}$ (mm/mm)	0	.000266	.000532	.0008
stress	80	100	120	140
strain	.001064	.00134	.0016	.00186
stress	160	180	190	200
strain	.00213	.003	.00468	.0100
stress	240	280	290	280
strain	.0260	.0680	.0880	.1680

The above results are plotted as a stress-strain graph in Figure 195.

Note that the elastic range (i.e. the straight line portion of the graph) for the material is very small. It is not possible to make accurate calculations based on this portion of the graph as drawn. To enable accurate

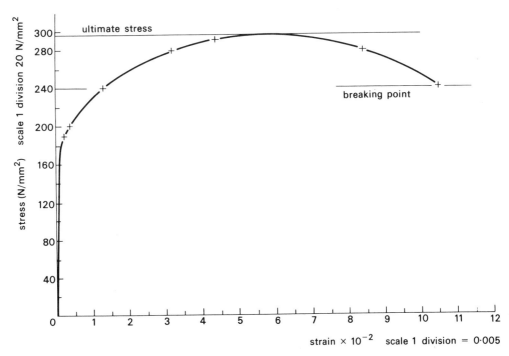

Figure 195   *Example 12*

calculations to be made, the results up to and including 20 kN are re-drawn to a larger scale in Figure 196. Using this graph it is now possible to make some accurate calculations.

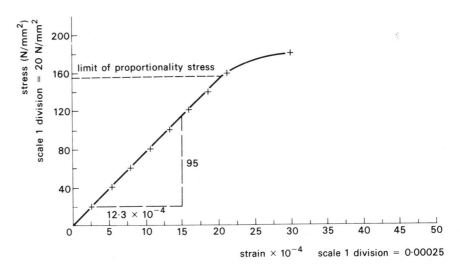

Figure 196   *Part stress-strain curve*

(i) *Modulus of elasticity*

From the straight line portion of the graph shown in Figure 196, using the slope of the graph:

$$E = \frac{95}{12.3 \times 10^{-4}} \text{ N/mm}^2 = 77.24 \text{ kN/mm}^2$$

or 77.24 GN/m^2

(ii) *Limit of proportionality stress*

From the graph the stress at the limit of proportionality is 156 N/mm^2.

(iii) *Ultimate stress*

The ultimate stress is the maximum stress value shown on the stress-strain curve. From the graph in Figure 195 the ultimate stress is 296 N/mm^2.

Percentage elongation and reduction in area

These two values can be calculated, or read directly from gauges, depending upon the type of testing equipment used. They give a guide to the ductility of the material being tested. This property will be discussed later in this topic area.

(iv) $\text{Percentage elongation} = \dfrac{\text{total extension}}{\text{original length}} \times 100$

For the specimen under test:

$$\text{percentage elongation} = \frac{10.5}{50} \times 100$$

$$= 21\%$$

(v) Percentage reduction in area

$$= \frac{\text{reduction in cross-sectional area}}{\text{original cross-sectional area}} \times 100$$

$$= \frac{(100 - 66.04)}{100} \times 100$$

$$= 33.96\%$$

The results from a standard tensile test enable some of the properties of a material to be established. These properties will now be discussed in further detail.

Ductility and brittleness

A ductile material is one which withstands large plastic deformations

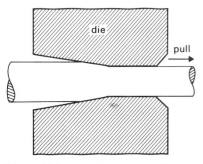

Figure 197 *A wire-drawing die*

without fracture. Conversely, a brittle material is one in which there is little, if any, plastic deformation before fracture occurs, and thus fracture usually occurs with very little warning.

The value of the percentage elongation and the percentage reduction in area give an indication of the ductility of the material. The higher these values, the greater is the extension per unit length of the material – i.e. the more stretch in the material. For example, high conductivity copper has a percentage elongation of 32% and a percentage reduction in area of 70%, whereas cast iron, a brittle material has a percentage elongation of 3% (see Figures 198 and 201).

Ductility is an important property of those materials which are subjected to cold-working processes such as drawing, cold-pressing and deep-drawing. Figure 197 shows a wire-drawing die. Drawing is exclusively a cold-working process, because it relies on the ductility of the material being drawn. Rod, wire and tubes are produced by drawing them through dies.

Cold-pressing and deep-drawing are two other manufacturing processes in which the ductility of the material used is important. In each of these processes the components are produced from sheet-stock, and range from pressed mild steel motor-car bodies to deep-drawn brass cartridge cases, cupro-nickel bullet envelopes, and aluminium milk churns. Only very ductile materials are suitable for deep-drawing. The best known of these are 70-30 brass (see Figure 199), pure copper, cupro-nickel, pure aluminium and some of its alloys, and some of the high nickel alloys.

Very many materials other than metals can be classified as either ductile or brittle. Glass for example, is a brittle material at room temperature, as are ceramics (bathroom and kitchen wall tiles for instance). Many polymers on the other hand, are ductile, but there are exceptions to this general rule – perspex is brittle at room temperature, but becomes ductile when heated.

Strength

The strength of a material is measured by the maximum stress it can withstand without fracturing. It is usual to express strength in terms of tensile strength. The value usually quoted is the ultimate strength; this is defined as the maximum applied force divided by the original cross-sectional area. For example, consider the two stress-strain curves shown in Figure 199. The curve for the material in the 'as drawn' condition shows an ultimate strength of 590 N/mm^2, and the curve for the annealed material shows an ultimate strength of 360 N/mm^2. The material in the as drawn condition has more strength than the material in the annealed state. Some typical values of tensile strengths are shown in the table.

material	typical tensile strengths $(N/mm^2 \text{ or } MN/m^2)$
high tensile engineering steel	1 500
commercial mild steel	450
high conductivity copper (hand drawn)	250
70/30 brass (as drawn)	590
70/30 brass (annealed)	360
aluminium bronze (as drawn)	1 650
stainless steel (EN56C air-hardened)	1 850
aluminium	59
lead	18
tin	11
zinc	110
European redwood	9
polyesters (setting types)	40
epoxy resins (general purpose)	63
alkyd resins	25
concrete	5
sintered aluminium power (10% aluminium oxide)	390
nylon 6	70–80

When considering the tensile strength of a material the ultimate strength is quoted. This value is calculated by using the equation:

$$\text{ultimate strength} = \frac{\text{maximum force}}{\text{original cross-sectional area}}$$

The actual breaking force, which may be less than the maximum force, is not used in calculations to give the tensile strength of a material. The value of the cross-sectional area at the instant when the specimen breaks is always less than the original cross-sectional area, so the true value of the breaking stress would have to be calculated using the smaller value for cross-sectional area. In practice the maximum force and the original cross-sectional area are used to calculate the ultimate strength of a material.

All the examples quoted to date have been for tensile strengths of materials. However, materials are also subjected to compressive and shearing forces. A material may have different values for its tensile, compressive and shear strengths. For example, the ultimate strength in tension of grey cast iron can vary from 154 to 400 N/mm^2. The corresponding range for the compressive strength of grey cast iron is from 600 up to 1 225 N/mm^2. The compressive strength of grey cast iron is therefore between three and four times the value of its tensile strength. Similarly the tensile strength of concrete is only one tenth of the value of its compressive strength.

Hardness

Hardness is that property of a material which measures its capacity to resist penetration or wear. Cutting tools need to be hard. Metal chutes or scrapers which are subject to abrasive wear need to have a hard surface.

There are a number of ways in which hardness is measured, the most common being by measuring indentation. A hardened steel ball of known diameter, or a diamond of known shape, is pressed by a known force into a metal specimen. The depth, or transverse dimension of the impression is a guide to the hardness of the specimen. There are a number of different hardness tests used, the most common being:

(i) Brinell Hardness Test to BSS 240,
(ii) Vickers Hardness Test to BSS 427,
(iii) Rockwell Hardness Test to BSS 891.

The hardness value of a material is expressed in terms of a number suffixed by the name of the system or machine used, e.g. 210 Brinell, 80 Rockwell scale C, 200 VPN. It is important to note that a hardness value of 200 Brinell is not the same as a hardness value of 200 VPN. Each method is independent of the others. For any of the three tests, the higher the number recorded the harder the material being measured. There is a close relationship between hardness and ultimate strength.

Comparison of stress-strain curves

The properties of a material depend upon several factors, two of the most important being (i) the composition of the material, and (ii) the structural condition of the material, e.g. cold drawn, annealed, normalized, etc.

Figures 198 to 203 show the stress-strain graphs for a range of different materials, together with the average mechanical properties and some typical uses for each material. Figure 198 shows the stress-strain graph for a cast iron. The low values for the percentage elongation and the absence of any reduction in area indicates that the material is brittle. Also note that there is no plastic range for this material.

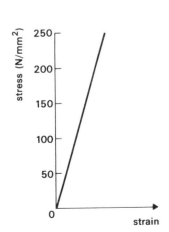

Material: cast iron — as cast

Average mechanical properties

Limit of proportionality stress	250 N/mm^2
Ultimate strength	250 N/mm^2
% Elongation	3%
% reduction in area	nil
Brinell hardness	180

Typical applications: Automotive engine blocks and castings for machine tools

Figure 198   *Cast iron stress-strain curve*

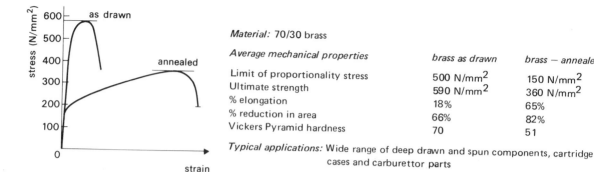

Material: 70/30 brass		
Average mechanical properties	*brass as drawn*	*brass — annealed*
Limit of proportionality stress	500 N/mm^2	150 N/mm^2
Ultimate strength	590 N/mm^2	360 N/mm^2
% elongation	18%	65%
% reduction in area	66%	82%
Vickers Pyramid hardness	70	51

Typical applications: Wide range of deep drawn and spun components, cartridge cases and carburettor parts

Figure 199   *70/30 brass stress-strain curve*

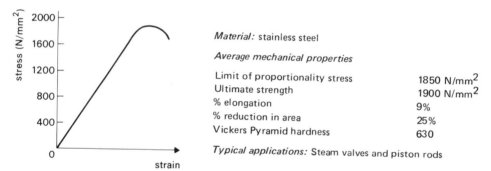

Material: stainless steel	
Average mechanical properties	
Limit of proportionality stress	1850 N/mm^2
Ultimate strength	1900 N/mm^2
% elongation	9%
% reduction in area	25%
Vickers Pyramid hardness	630

Typical applications: Steam valves and piston rods

Figure 200   *Stainless steel stress-strain curve*

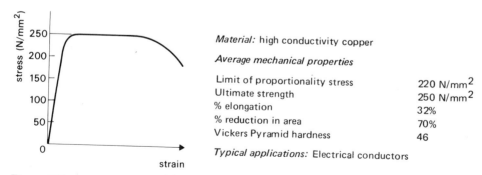

Material: high conductivity copper	
Average mechanical properties	
Limit of proportionality stress	220 N/mm^2
Ultimate strength	250 N/mm^2
% elongation	32%
% reduction in area	70%
Vickers Pyramid hardness	46

Typical applications: Electrical conductors

Figure 201   *High conductivity copper stress-strain curve*

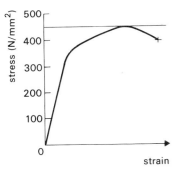

Figure 202   *Aluminium alloy*
stress-strain curve

Material: aluminium alloy

Average mechanical properties

Limit of proportionality stress	340 N/mm^2
Ultimate strength	450 N/mm^2
% elongation	23%
% reduction in area	40%
Vickers Pyramid hardness	155

Typical applications: Duralumin alloy — high strength to weight ratio, used in aircraft industry

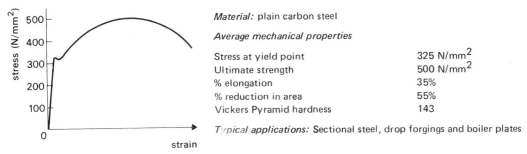

Figure 203   *Plain carbon steel stress-strain curve*

Material: plain carbon steel

Average mechanical properties

Stress at yield point	325 N/mm^2
Ultimate strength	500 N/mm^2
% elongation	35%
% reduction in area	55%
Vickers Pyramid hardness	143

Typical applications: Sectional steel, drop forgings and boiler plates

Figure 199 shows two stress-strain graphs for a 70/30 brass. A 70/30 brass is a copper-base alloy containing 30% zinc. These graphs illustrate how the properties of a material can be changed by heat treatment. The material in the 'as drawn' condition has been passed through a die, the cross-section has been reduced and the material has been work hardened, resulting in an increase in the strength and a decrease in the ductility. If this material is then annealed, the material returns to its original ductile state. The increase in ductility is shown by the increase in the percentage elongation from 18% to 65%, and the much higher value of strain at fracture of the annealed material.

Figure 200 shows the stress-strain curve for a stainless steel. This material is a strong material because of its high ultimate strength value of 1 900 N/mm^2 (also note the high hardness value). It is not very ductile as illustrated by the low percentage elongation value of 9%.

Figure 201 shows the stress-strain curve for a high conductivity copper.

This material is not very strong as indicated by the ultimate strength value (note the low hardness value), but it is ductile as indicated by the percentage elongation value.

Figure 202 shows the stress-strain curve for an aluminium alloy. This material has strength and ductility values approximately the same as those of a plain carbon steel (see Figure 203). It is, however, much lighter than steel, and as a consequence is used extensively in the air-craft industry. However, although aluminium is much lighter than plain carbon steel it is much more expensive.

Figure 203 shows the stress-strain curve for a low carbon steel (up to about 0.3% carbon). There is a clear distinction in shape between this curve and the preceding curves — at the yield point there is an increase in strain without any increase in stress. In other words the material yields.

Apart from the yield point, the stress-strain curve for low carbon steels is similar to those of other ductile materials. The material is strong and stiff, relatively cheap, and therefore used extensively in industry.

Summary

A ductile material is one which will withstand large plastic deforma-tions without fracture. The value of percentage elongation and the percentage reduction in area give an indication to the ductility of a material. When comparing stress-strain graphs drawn to the same scale as shown in Figure 204, the greater the strain value at fracture the more ductile the material.

The strength of a material can be measured by the value of the stress it can withstand without fracture occurring. The tensile strength of a material is the ultimate strength taken from the stress-strain graph. When comparing stress-strain graphs drawn to the same scale as shown in Figure 204, the greater the ultimate strength value the stronger the material. Remember the tensile strength of a material may be different from its compressive or shear strength.

When comparing stress-strain graphs drawn to the same scale as shown in Figure 204, the steeper the slope of the straight line portion of the graph the stiffer the material.

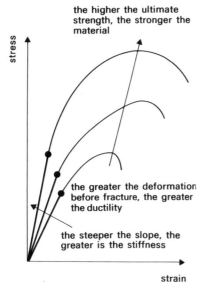

the higher the ultimate strength, the stronger the material

the greater the deformation before fracture, the greater the ductility

the steeper the slope, the greater is the stiffness

Figure 204   *Ductility, strength and stiffness of materials*

Self-assessment questions

19 Define each of the following terms with respect to metals used in engineering:
(*a*) Ductility.
(*b*) Brittleness.
(*c*) Strength.
(*d*) Hardness.
(*e*) Limit of proportionality.
(*f*) Elastic limit.

20 Describe the procedure to be carried out in order to perform a tensile test to destruction. Give details of all the results that should be recorded.

21 The following results were obtained from a tensile test on a low carbon steel. The specimen used had a cross-sectional area of 100 mm² and a gauge length of 50 mm.

force (kN)		4	8	12	16	20	24	28	32
extension (mm)		0	.0020	.012	.02	.03	.035	.049	.056

force (kN)	34	36	38	40	42	46	60	80	90
extension (mm)	.065	.068	.072	.077	.081	.089	0.9	3	5

yield force   = 47 kN   final diameter = 9.74 mm
maximum force = 92 kN   final length   = 61.56 mm
breaking force = 80 kN

Using the above results:
(*a*) (i) plot a stress-strain curve for values up to the limit of proportionality,
(ii) plot a stress-strain curve for the complete range of results.

(*b*) Calculate for the material:
(i) the modulus of elasticity,
(ii) the yield stress,
(iii) the ultimate strength,
(iv) the percentage elongation,
(v) the percentage reduction in area.

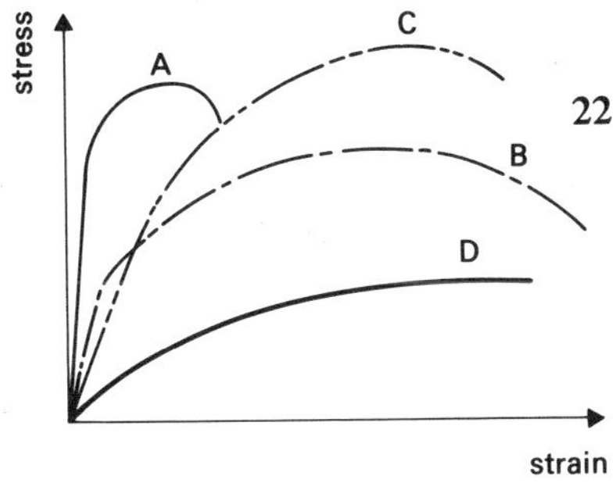

22 Figure 205 shows the stress-strain curves for four different materials, labelled A, B, C and D. The four curves are drawn to the same scales. Study the four curves and then answer the following questions giving reasons for your answers.
(i) Which material has the highest value of ultimate strength?
(ii) Which material is the most ductile?
(iii) Which material has the highest limit of proportionality stress?
(iv) Which material has the highest value of Young's modulus?
(v) Which material is the most flexible (i.e. the least stiff)?
(vi) Which material does not obey Hooke's law?

Figure 205   *Self-assessment question 22*

Solutions to self-assessment questions

19　(a)　Ductility – the ability to withstand large plastic deformations witho[u]
fracture.

(b)　Brittleness – the inability to withstand large plastic deformations – fractu[re]
may occur with very little warning.

(c)　Strength – the ability to withstand large stresses without fracture.

(d)　Hardness – the ability to withstand surface indentation.

(e)　Limit of proportionality – the point where stress ceases to be proportion[al]
to strain.

(f)　Elastic limit – up to the elastic limit a material returns to its original shap[e]
when applied forces are removed – beyond this point the material deform[s]
permanently.

20　The procedure should be:

(1)　Measure initial diameter and gauge length – record these.

(2)　Set up specimen and extensometer – check zero readings for force an[d]
extension.

(3)　Load specimen – record force and extension values.

(4)　Remove specimen after fracture – take final diameter and extension readings[.]

21　The stress-strain values are as shown below:

		4	8	12	16	20	24	28	32
force, F (kN)	=	4	8	12	16	20	24	28	32
stress = F/A	=	40	80	120	160	200	240	280	320
$= \dfrac{F \times 1000}{100}$ (N/mm²)									
extension (mm)	=	0	.002	.012	.02	.03	.035	.049	.056
strain = x/ℓ	=	0	.004	.024	0.4	0.6	0.7	0.98	1.12
$= (x/50) \times 10^{-3}$									

force	=	34	36	38	40	42	46	60	80	90
stress	=	340	360	380	400	420	460	600	800	900
extension	=	.065	.068	.072	.077	.081	.089	.9	3	5
strain $\times 10^{-3}$	=	1.3	1.36	1.44	1.54	1.62	1.78	18	60	100

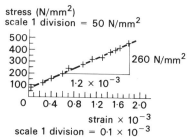

stress (N/mm²)
scale 1 division = 50 N/mm²

260 N/mm²

1.2×10^{-3}

strain $\times 10^{-3}$
scale 1 division = 0.1×10^{-3}

Figure 206　*Stress-strain curve*
up to limit of proportionality

(a)　(i)　The graph of stress-strain up to the limit of proportionality is as shown in
Figure 206.

(ii)　The graph for the complete range of results is as shown in Figure 207.

(b)　(i)　Using values from the graph shown in Figure 206,

$$E = \frac{260}{1.2 \times 10^{-3}} \quad \frac{N}{mm^2}$$

$$E = 216.7 \text{ kN/mm}^2 \text{ or } 216.7 \text{ GN/m}^2$$

(ii)　Yield stress $= \dfrac{\text{force at yield point}}{\text{original cross-sectional area}}$

$$= \frac{47 \times 10^3}{100} \frac{[N]}{[mm^2]} = 470 \text{ N/mm}^2$$

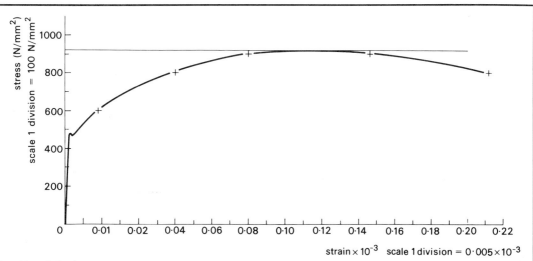

Figure 207   *Complete stress-strain curve*

   (iii) Ultimate strength

$$= \frac{\text{maximum force}}{\text{original cross-sectional area}}$$

$$= \frac{92 \times 10^3}{100} \frac{[\text{N}]}{[\text{mm}^2]} = 920 \text{ N/mm}^2$$

   (iv) % elongation $= \dfrac{\text{total extension}}{\text{original length}} \times 100\%$

$$= \frac{(61.56 - 50)}{50} \times 100 = 23.12\%$$

   (v)   % reduction in area

$$= \frac{\text{reduction in cross-sectional area}}{\text{original cross-sectional area}} \times 100\%$$

$$= \frac{(100 - 74.5)}{100} \times 100 = 25.5\%$$

22   (i)   Material C – because it has the highest value on the stress scale.
   (ii)   Material B – because it has the largest strain – i.e. it stretches farthest along the strain axis.
   (iii)   Material A – because it has the highest value of stress at the end of the straight line portion.
   (iv)   Material A – because it has the largest slope of the straight line portion.
   (v)   Of the three materials which have a straight line portion – the material C has the smallest value of E and is the most flexible. However, the material D is the most flexible of the four materials.
   (v)   Material D because there is no straight line portion to the graph.

Section 2
Co-planar forces

After reading the following material, the reader shall:

2   Determine the resultant of a number of forces acting at a point.

2.1   State the two conditions for equilibrium.

2.2   Define resultant force.

2.3   Define equilibrant force.

2.4   Determine the resultant of two co-planar forces by drawing, giving magnitude and direction.

The reader will, in previous study, have dealt with the concept of equilibrium. However, an understanding of this concept is so important to the following part of this topic area that the whole idea of equilibrium will be reviewed.

When two or more forces act upon a body and are so arranged that the the body remains at rest or moves with a constant speed in a straight line, the forces and the body are in equilibrium. The emphasis on constant speed in a straight line is important because a body which is accelerating (i.e. a body whose velocity is changing) is subjected to an accelerating force, and is therefore not in a state of equilibrium.

The simplest case of equilibrium is that of two equal and opposite forces acting in the same straight line. The forces may be two equal and opposite pulls, as shown in Figure 208 (*a*), in which case the body is said to be in tension. If there are only two forces in equilibrium these two forces must be equal and opposite, and further, must act in the same straight line, as shown in Figure 208 (*a*). If the two forces are equal in magnitude but not acting in the same straight line, the body will rotate (if possible) until the two lines of action are in the same straight line as shown in Figure 208 (*b*).

If a body is not to move laterally, all the forces acting on it must balance, i.e. the body is in equilibrium. Figure 209 (*a*) illustrates a very simple example. In this case the horizontal forces have a zero resultant, as do the vertical forces. If the forces are inclined to these two directions as shown in Figure 209 (*b*), then it is possible to check to see if the body is in equilibrium by either resolving the forces into horizontal and vertical components and adding all the components, or by adding the vectors. This latter case will be considered in detail in this topic area. In any case *if the body is in equilibrium then the vector sum of the forces must be zero*. This is one of the two conditions for equilibrium.

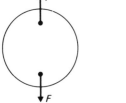

two equal and opposite forces

(*a*)

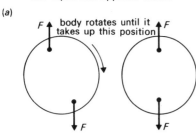

two equal and opposite forces
not in the same straight line

(*b*)

Figure 208   *Equal and opposite forces*

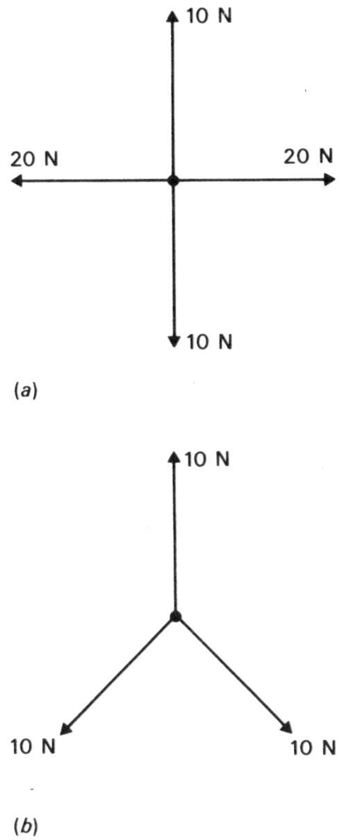

10 N

20 N   20 N

10 N

(a)

10 N

10 N   10 N

(b)

Figure 209   *Forces acting on a body*

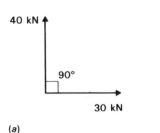

40 kN

90°

30 kN

(a)

40 kN

predicted direction of resultant force

30 kN

(b)

Figure 210   *Example 13*

The second condition for equilibrium is that there should be no resultant turning moment acting on the body. The clockwise turning moments acting on the body must balance the anti-clockwise turning moments acting on the body. *That is, the algebraic sum of the moments of all the forces about any point in the plane must be zero.*

The reader will already be familiar with the concept of force as a vector quantity. A force has magnitude and direction, and therefore can be represented by a vector. Consideration will be given only to *forces acting in one plane*, these forces are referred to as *co-planar forces*.

Resultant and equilibrant forces

Example 13
Determine the resultant of the two co-planar forces shown in Figure 210.

The *resultant* of a system of forces is that single force which has the same effect on the body as the original system of forces.

The magnitude and direction of a resultant force is often found by using a vector diagram. However, from an inspection of Figure 210 (*a*), it may be appreciated that if two forces act as shown, then the resultant of these forces must lie somewhere between them. It may also be anticipated that the resultant force will lie nearer to the 40 kN force than to the 30 kN force. These predictions are summarized in Figure 210 (*b*).

The diagram in Figure 210 is called the *space diagram*. To determine the resultant of these two forces it is necessary to add vectorially the two forces together. This is carried out as follows:

1   Draw the vector **ab** to represent the 40 kN force, as shown in Figure 211 (*a*).
2   On the end of the vector **ab**, i.e. at *b*, draw the vector **bc** to represent the 30 kN force, as shown in Figure 211 (*b*).
3   The resultant of these two forces is represented by the vector **ac**.

Note that the arrow on vector **ac** corresponds to the predicted direction of the resultant on Figure 210 (*b*). The arrows on vectors **ab** and **bc** follow sequentially, nose-to-tail. However, the arrow on the vector **ac** does not follow this order. The arrow is in the opposite direction.

The magnitude of the resultant force can be scaled from the vector or force diagram shown in Figure 211 (*c*); the angular position can be measured with a protractor. For this example the resultant force is 50 kN at 53° 8' as shown in Figure 211 (*c*).

For a body to be in equilibrium, the vector sum of the forces acting on the body must be zero and the algebraic sum of the moments must be zero. In many cases a body which is acted upon by a number of

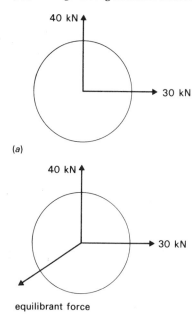

(a)

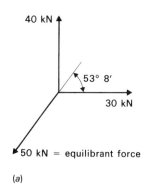

equilibrant force

(b)

Figure 212   *Equilibrant force*

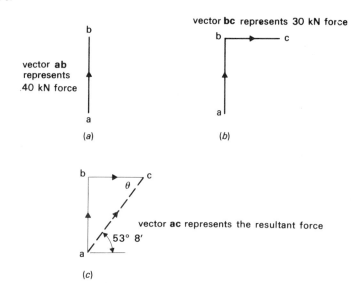

Figure 211   *Vector diagram*

forces may be brought into a state of equilibrium by the addition of a balancing force. Such a balancing force is called an equilibrant force. Again using Example 13 to illustrate the point, a two force system is shown in Figure 212 (a), and it is not difficult to visualize that if some third force were added as shown in Figure 212 (b), the system would then be in equilibrium. What then is the magnitude and direction of this particular equilibrant force? How, as a general rule, can the equilibrant force be obtained for a system of two or more inclined forces? *The equilibrant force is equal in magnitude but opposite in direction to the resultant force.*

Thus for example 13, the equilibrant force is as shown in Figure 213. To obtain this equilibrant force for a system of forces the procedure is to draw the force diagram (as detailed earlier and as shown in Figure 211); the equilibrant force is represented by the vector **ca**, i.e. in the direction opposite to the resultant force which is represented by vector **ac**.

Figure 213 (b) shows the vector diagram for the three forces. Because the vector diagram closes, there is no resultant force, and hence the forces are in equilibrium. Note that when the forces are in equilibrium the arrows on the vectors follow round nose to tail. Consider the following examples:

Example 14
Figure 214 shows the supports of a lathe steady with the forces acting on them. Find the resultant of these two forces and the angle at which it acts in relation to the horizontal support. What is the magnitude and direction of the equilibrant force?

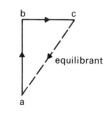

(a)

(b)

Figure 213   *Forces in equilibrium*

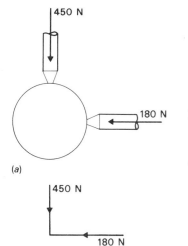

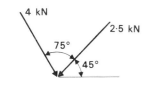

space diagram – showing forces acting

(b)

Figure 214 *Example 14*

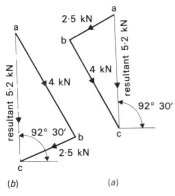

Figure 216 *Example 15*

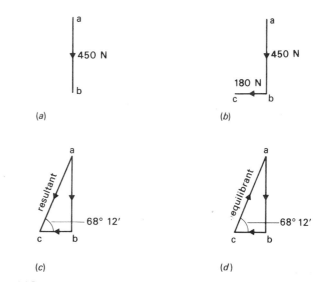

(c) *(d)*

Figure 215 *Force vector diagram*

(b) *(a)*

Figure 217 *Resultant of perpendicular forces*

To determine the resultant and equilibrant forces follow the procedure described earlier, that is:

1 Draw the vector **ab** to represent one of the forces, say 450 N, as shown in Figure 215 *(a)*.

2 Add to the vector **ab** the vector **bc** to represent the force of 180 N, as shown in Figure 215 *(b)*.

3 To determine the resultant force, complete the force diagram by drawing the vector **ac**, as shown in Figure 215 *(c)*. The resultant force is 484.7 N at 68° 12′ to the horizontal.

4 The equilibrant force is equal in magnitude to the resultant force, but is opposite in direction. Therefore the equilibrant force is represented by the vector **ca**, as shown in Figure 215 *(d)*. Note that the arrows on the vectors follow round nose to tail when the forces are in equilibrium.

Example 15

Determine the value of the resultant force of the two forces shown in Figure 216.

The procedure for solving the problem is the same as described in Examples 13 and 14.

Draw the vector **ab** to represent the force of 2.5 kN.

To this vector, add the vector **bc** to represent the force of 4 kN (see Figure 217 *(a)*).

The resultant force is given by vector **ac** as shown in Figure 217 *(a)*, i.e. 5.2 kN at 92° 30′ to the horizontal.

Figure 217 *(b)* shows the same example solved by taking the 4 kN force first and representing it by vector **ab**. The answer is the same in either case. It does not matter which force is considered first.

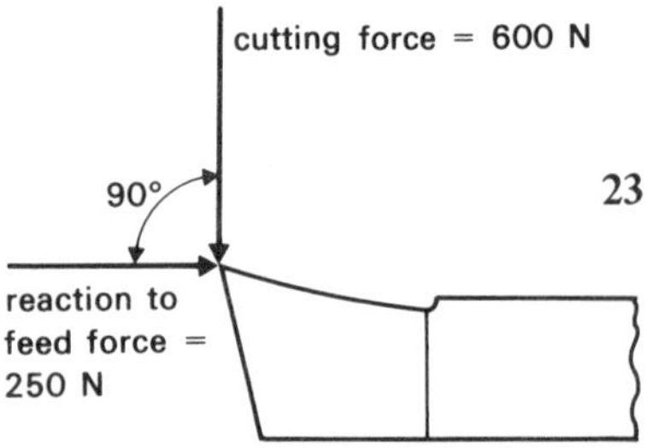

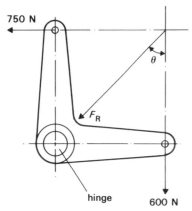

Figure 218   *Self-assessment question 25*

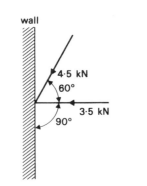

Figure 219   *Self-assessment question 26*

Figure 220   *Self-assessment question 27*

Self-assessment questions

Study each of the following statements and indicate whether they are true or false.

23 (i) A body which is accelerating is in a state of equilibrium.

TRUE/FALSE

(ii) A body which is moving at a constant speed in a straight line is in a state of equilibrium.

TRUE/FALSE

(iii) For a body to be in a state of equilibrium it is necessary only for the vector sum of the forces acting on it to be zero.

TRUE/FALSE

(iv) Co-planar forces are forces acting in a single plane only.

TRUE/FALSE

(v) A resultant force is a single force which can replace two or more forces.

TRUE/FALSE

(vi) If two or more forces are replaced by a resultant force the effect on the body is changed.

TRUE/FALSE

(vii) An equilibrant force is the force which, if applied to a body, will cause the body to be in a state of equilibrium.

TRUE/FALSE

(viii) The equilibrant force is identical to the resultant force.

TRUE/FALSE

24 State the two conditions necessary for a body to be in equilibrium.

25 Figure 218 shows a side elevation of a parting off tool. The cutting force is 600 N, whilst the 250 N reaction is caused by feeding the tool into the metal. Find the resultant force acting on the tool's cutting edge, and the angle at which it acts relative to the 250 N force.

26 Figure 219 shows a cranked lever which is part of a gear change mechanism. Find the resultant force F_R acting on the hinge pin and the angle θ.

27 Figure 220 shows two members of a frame structure which meet at a wall. Find the magnitude and direction of the equilibrant force at the wall.

After reading the following material, the reader shall:

2.5 Use Bow's notation.

2.6 Use the polygon of forces to solve problems involving more than two co-planar forces.

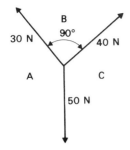

Figure 221   *Application of Bow's notation*

The logical continuation from the consideration of two co-planar forces is to consider more than two forces. The procedure for dealing with problems involving any number of forces is similar to that for dealing with two forces; however before proceeding to the solution of problems, it is necessary to discuss a method of identifying the different forces.

Bow's notation

Consider the three forces shown in Figure 221. The lines of action of the three forces pass through one point on the space diagram. On the space diagram the spaces between the forces are lettered using capital letters, as indicated in Figure 221. The 30 N force can then be referred to as force AB, the 40 N force can be referred to as force BC, and the 50 N force can be referred to as force CA. This method of lettering a space diagram is known as Bow's notation.

Figure 222 shows three space diagrams lettered according to Bow's notation. The lettering can start in any space and continue in any direction, but all space diagrams in this text will be lettered in a clockwise direction.

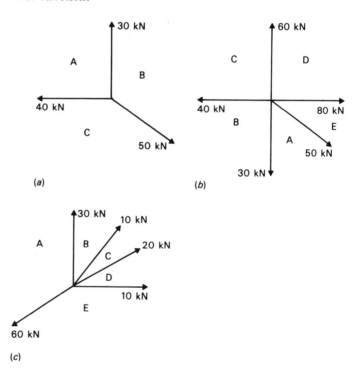

Figure 222   *Bow's notation*

Before proceeding to solve problems involving more than two forces, consider again the two conditions necessary for equilibrium. (i) The vector sum of the forces acting on a body must be zero. For this to be true then there must be no resultant force acting on the body; therefore the force diagram closes. If the force diagram does not close, there is a resultant force acting on the body, and an equilibrant force is required to bring the system of forces into a state of equilibrium. (ii) The second condition required for equilibrium is that the algebraic sum of the moments of all the forces about any point in the plane must be zero.

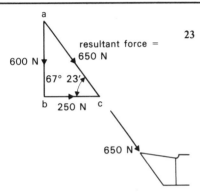

Figure 223   *Solution to self-assessment question 25*

Solutions to self-assessment questions

23 (i)   FALSE – for a body to be in a state of equilibrium there can be no unbalanced forces (such as an accelerating force) acting on the body.

(ii)   TRUE – a body at rest or moving in a straight line at a constant speed is in a state of equilibrium.

(iii)   FALSE – the vector sum of the forces and the algebraic sum of the moments about any point must be zero for a body to be in equilibrium.

(iv)   TRUE – 'co-planar' means single plane.

(v)   TRUE.

(vi)   FALSE – a resultant force has the same effect as the separate forces which it replaces.

(vii)   TRUE.

(viii)   FALSE – the equilibrant force is equal in magnitude but opposite in direction to the resultant force.

24 (i)   The vector sum of the forces acting on a body must be zero.

(ii)   The algebraic sum of the moments of all the forces about any point in the plane must be zero.

25   The force diagram is shown in Figure 223.
Resultant force = 650 N at 67° 23′ to horizontal.
[Note that when a resultant force is drawn, the arrows *do not* follow nose to tail.]

26   The force diagram is shown in Figure 224.
Resultant force = 960 N at 51° 20′.
[Note that when a resultant force is drawn, the arrows *do not* follow nose to tail.]

27   The force diagram is shown in Figure 225.
Equilibrant force = 7 kN at 35° to horizontal.
[Note that when an equilibrant force is drawn, the arrows *do* follow nose to tail.]

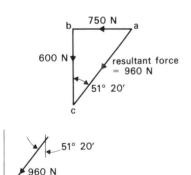

Figure 224   *Solution to self-assessment question 26*

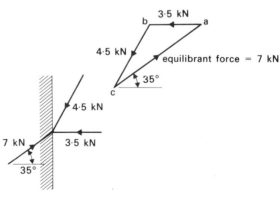

Figure 225   *Solution to self-assessment question 27*

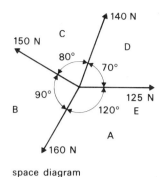

space diagram

Figure 226 *Example 16*

The following examples demonstrate the solution of problems involving more than two forces which are co-planar (i.e. act in one plane), and concurrent (i.e. pass through the same point). It should be noted that Bow's notation is widely used in the solution of problems involving (i) forces is several planes and (ii) forces which do not act through one point.

Example 16

Find the magnitude and direction of the equilibrant and resultant forces of the system shown in Figure 226.

Letter the space diagram according to Bow's notation, as shown in Figure 226. Note that letters A and E appear to occupy the same space, the reason for this will be discussed later.

Draw the vector **ab** to represent the force AB as shown in Figure 227 (*a*). This vector is drawn parallel to the line of action of force AB.

On the end of vector **ab**, i.e. from point *b*, draw the vector **bc** to represent the force BC. This vector is drawn through *b* and parallel to the line of action of force BC as shown in Figure 227 (*b*).

On the end of vector **bc**, i.e. from point *c*, draw the vector **cd** to represent the force CD. This vector is drawn through *c* and parallel to the line of action of force CD as shown in Figure 227 (*c*).

On the end of vector **cd**, i.e. from point *d*, draw the vector **de** to represent the force DE. This vector is drawn through *d* and parallel to the line of action of force DE as shown in Figure 227 (*d*).

If the forces are in equilibrium, the point *e* on the vector diagram must coincide with point *a*, i.e. the diagram must close, and the vector sum of the forces is zero. When the vector diagram does not close, this means that there is a resultant force, and that an equilibrant force is required to bring the system of forces into a state of equilibrium.

For the example under consideration the forces are not in equilibrium, and this is why the letter E has to be included on the space diagram. This allows for the resultant force represented by vector **ae** and the equilibrant force represented by vector **ea** to be scaled from the vector diagram. The magnitude of the resultant and equilibrant forces is 78 N and the directions are as shown on Figure 227 (*e*).

In all examples where a resultant or equilibrant force is anticipated, it is necessary to include an extra letter as has been done in Example 16.

A vector diagram can be used to determine unknown forces in a system of forces which is in equilibrium:

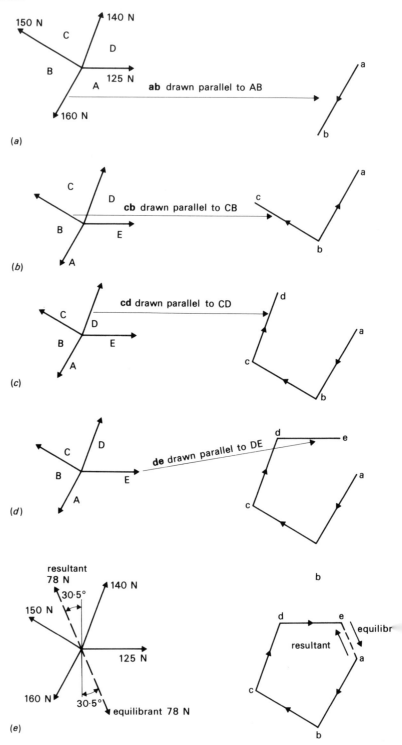

Figure 227   *Equilibrant and resultant using Bow's notation*

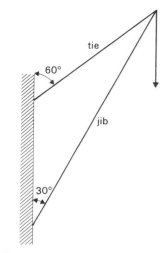

Figure 228   *Example 17*

Example 17

Figure 228 shows in diagrammatic form a small crane which swings about a vertical spindle. The maximum safe load that can be carried is 4 tonnes. Find the forces in the jib and tie of the crane, when it is carrying the maximum load.

There are three forces acting at the one point. These are the force due to the 4 tonne mass, the force in the jib and the force in the tie. What is known about these three forces? For the force due to the 4 tonne mass the magnitude and direction is known. For the other two forces only the directions are known; it is required to find the magnitude of these forces.

These two values can be determined by the following method. Figure 235 (*a*) shows the space diagram, lettered using Bow's notation. Because both the magnitude and direction of the force AB are known, the vector **ab** to represent this force can be drawn as shown in Figure 235 (*b*).

The next force to be considered is force BC. Because only the direction of this force is known, it is not possible to draw the complete vector **bc**. However, it is possible to draw a line through point *b*, parallel to the line of action of the force BC, as shown in Figure 235 (*c*).

Consider now the force CA. Again, because only the direction of this force is known, it is not possible to draw the complete vector **ca**. However, it is possible to draw a line through point *a*, parallel to the line of action of the force CA, as shown in Figure 235 (*d*).

The two vectors **bc** and **ca** have a common point in *c*. This point is given by the intersection of the two lines drawn parallel to the lines of action of the forces BC and CA. The complete force diagram is shown in Figure 235 (*e*). The magnitude of the forces BC and CA can be scaled from the vector diagram.

Thus, the force in the jib, i.e. force BC = 69 kN,
and   the force in the tie, i.e. force CA = 40 kN

Self-assessment questions

28 Figure 229 shows a wall crane which can carry a maximum safe load to 2 tonnes. Determine the forces in the jib and tie when the crane carries the maximum load of 2 tonnes.

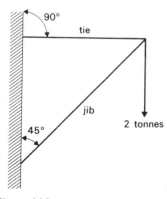

Figure 229   *Self-assessment question 28*

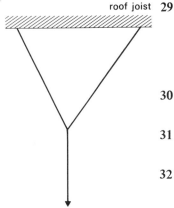

roof joist

2 tonnes

Figure 230   *Self-assessment question 29*

29 Two wire ropes, 1.25 m and 1.75 m long, are suspended from two supports, 1.75 m apart, fixed to a roof joist as shown in Figure 230. The lower ends of the ropes are brought together and a pair of chain blocks hung on them. If the blocks are used to lift a mass of 2 tonnes determine the force in each rope.

30 Find the magnitude and direction of the equilibrant and resultant forces for the system shown in Figure 231.

31 Figure 232 shows the forces acting on a bearing. Determine the resultant force on the bearing.

32 An irregularly shaped casting clamped to the face-plate of a centre lathe causes out-of-balance forces on the rotating lathe spindle. The magnitudes and angular positions of these forces are as shown in Figure 233. Determine the magnitude and direction of the force required to balance the system.

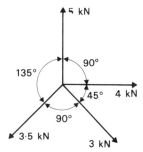

Figure 231   *Self-assessment question 30*

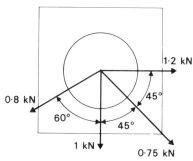

Figure 232   *Self-assessment question 31*

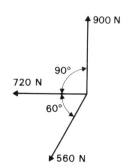

Figure 233   *Self-assessment question 32*

Solution to self-assessment question

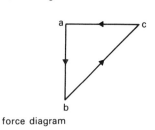

space diagram

force diagram

28 The force diagram is shown in Figure 235. From this diagram the force in the jib is 27.75 kN and the force in the tie is 19.62 kN.

Figure 234   *Solution to self-assessment question 28*

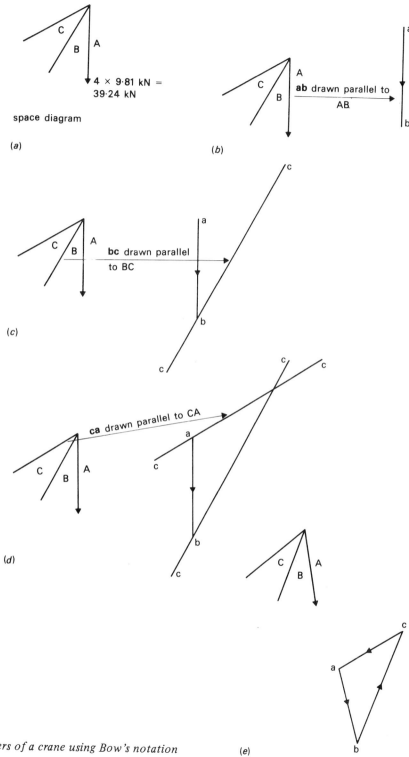

Figure 235 *Forces in the members of a crane using Bow's notation*

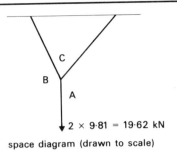

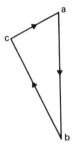

space diagram (drawn to scale)

force diagram

Solutions to self-assessment questions

29 The force diagram is shown in Figure 236. From this diagram the forces in each of the ropes are 7.2 kN and 15.6 kN.

30 The force diagram is shown in Figure 237. From this diagram the resultant force is 3.7 kN (i.e. vector **ae**) at 6.5° to the horizontal, and the equilibrant force is 3.7 kN (i.e. vector **ea**) at 6.5° to the horizontal.

31 The force diagram is shown in Figure 238. From this diagram the resultant force is 2.2 kN at 62° to the horizontal.

32 The force diagram is shown in Figure 239. From this diagram the equilibrant force is 1080 N at 22° below the horizontal.

Figure 236   *Solution to self-assessment question 29*

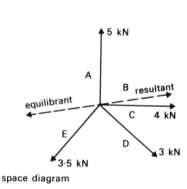

space diagram

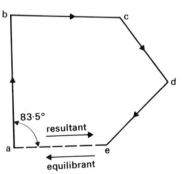

Figure 237   *Solution to self-assessment question 30*

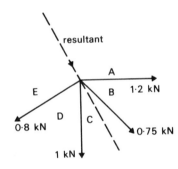

space diagram

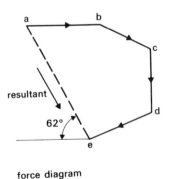

force diagram

Figure 238   *Solution to self-assessment question 31*

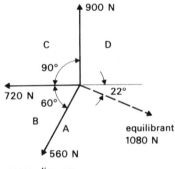

space diagram

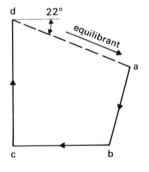

force diagram

Figure 239   *Solution to self-assessment question 32*

Section 3
Simply supported beams

After reading the following material, the reader shall:

3   Solve problems of simply supported beams.
3.1 State that upward forces are equal to downward forces at equilibrium.
3.2 Determine the reaction at the supports for a uniform simply supported beam having point loads.

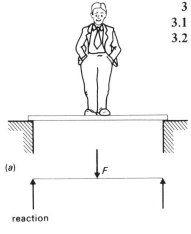

(a)

reaction

(b)

Figure 240   *Simply supported beam carrying a point load*

In engineering practice it is not unusual to have a structure, or a member of a structure, which is subjected to a system of forces having parallel lines of action. To illustrate this type of loading consider the very simple example illustrated in Figure 240. Figure 240 (*a*) illustrates a person standing at the centre of a plank of timber which spans a hole in the ground, the plank being supported on either side of the hole. The forces involved in this simple example are shown in Figure 240 (*b*). The force due to the mass of the person is shown by an arrow acting vertically downwards. The forces exerted by the ground on the plank are shown by arrows acting vertically upwards. These upward forces are referred to as *reactions*, the load on the beam is referred to as a *point load* or *concentrated load* and the plank supporting structure is referred to as a *beam*. The whole is referred to as a *simply supported beam*, subject to a point load. The type of support is called simply supported because it is assumed that the beam is supported at a single point or on a knife edge, and not constrained in any way.

Before considering any problems involving simply supported beams carrying point loads, it is advisable to re-state the two conditions required for equilibrium. The conditions are:

The vector sum of the forces must be zero,
and
The sum of the moments about any point must be zero.

For a simply supported beam subjected to vertical point loads, the first condition for equilibrium can be re-stated as upward forces must equal downward forces.

In all the examples considered, the effect of the mass of the beam will be neglected. Very often in practice this cannot be ignored — just think of the mass of steelwork in an overhead crane main girder, or in a bridge. However, at this stage of study, the effect of the mass of the beam is an unnecessary complication.

In problems involving simply supported beams, it is necessary to take

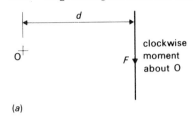

(a)

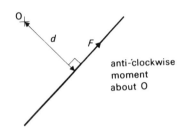

(b)

Figure 241   *Moment of a force*

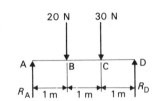

Figure 242   *Example 18*

moments of forces about given points. The reader should know that:

moment of a   =   force $\times$ perpendicular distance between
force about an   the line of action of the force and
axis   the axis.

Referring to Figure 241, the moment of force F about the axis through point O is the product of F and d. The basic unit for the moment of a force is the newton-metre (Nm). The moment can act in a clockwise direction as shown in Figure 241 (*a*), or in an anticlockwise direction as shown in Figure 241 (*b*).

The condition for equilibrium that the sum of the moments about any point must be zero can be re-stated for use in simple beam problems and many other problems as:

clockwise moments = anticlockwise moments

Consider the following examples:

Example 18

A simply supported beam is loaded as shown in Figure 242. Determine the magnitude of the two reaction forces at A and D.

Let the reaction force at A be R_A and the reaction force at D be R_D. The system of forces has two unknowns, R_A and R_D. Take moments about an axis through A. The force at A then produces no moment:

anticlockwise moments = $R_D[N] \times 3[m] + R_A[N] \times 0[m]$ = $3R_D$

clockwise moments   = $20[N] \times 1[m] + 30[N] \times 2[m]$
                        = $20 + 60$ Nm
                        = 80 Nm

For equilibrium:
clockwise moments   = anticlockwise moments
                   $\therefore$ 80 = $3R_D$

$$R_D = \frac{80}{3}\frac{[Nm]}{[m]} = 26.67 \text{ N}$$

Taking moments about an axis through D, the force R_D produces no moment:

clockwise moments   = $R_D[N] \times 0[m] + R_A[N] \times 3[m]$ = $3R_A$

anticlockwise moments = $30[N] \times 1[m] + 20[N] \times 2[m]$
                        = $30 + 40$ Nm
                        = 70 Nm

Equating clockwise and anticlockwise moments:
$$3R_A = 70$$

$$\therefore \quad R_A = \frac{70 \,[\text{Nm}]}{3 \,[\text{m}]}$$

$$R_A = 23.33 \,\text{N}$$

To confirm the arithmetic, check that the beam fulfils the condition regarding forces in equilibrium.

i.e. upward forces $= R_A + R_D = 23.33 + 26.67 \quad = 50 \,\text{N}$
downward forces $= 20 + 30 \qquad\qquad\qquad\qquad = 50 \,\text{N}$
therefore, upward forces = downward forces.
Hence the values of R_A and R_D are correct.

The reader could have noticed that, once R_D has been determined by taking moments, the reaction at A could be found by considering the vertical equilibrium of the beam, instead of taking moments for a second time; the amount of calculation involved may then be reduced. However, until the reader is sufficiently well practised, it is recommended that he use the method suggested above, and use the consideration of the equilibrium of vertical forces as a check on the accuracy of his work.

Consider a rather more complex example.

Example 19
A horizontal beam, A, B, C, D, E is loaded as shown in Figure 243. Determine the reaction forces at A and D.

Let R_A and R_D be the reaction forces at A and D respectively. Take moments about an axis through A.

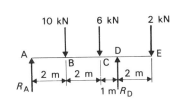

Figure 243   *Example 19*

anticlockwise moments $= R_A\,[\text{kN}] \times 0\,[\text{m}] + R_D\,[\text{kN}] \times 5\,[\text{m}]$
$\qquad\qquad\qquad\quad = 5R_D \,\text{kNm}$

clockwise moments $\quad = 10\,[\text{kN}] \times 2\,[\text{m}] + 6\,[\text{kN}] \times 4\,[\text{m}]$
$\qquad\qquad\qquad\qquad\qquad + 2\,[\text{kN}] \times 7\,[\text{m}]$
$\qquad\qquad\qquad\quad = 20 + 24 + 14 \,\text{kNm}$
$\qquad\qquad\qquad\quad = 58 \,\text{kNm}$

Equating clockwise and anticlockwise moments:
$$5R_D = 58$$

$$R_D = \frac{58\,[\text{kNm}]}{5\,[\text{m}]}$$

$$R_D = 11.6 \,\text{kN}$$

Take moments about an axis through D

clockwise moments $\quad = R_D\,[\text{kN}] \times 0\,[\text{m}] + R_A\,[\text{kN}] \times 5\,[\text{m}]$
$\qquad\qquad\qquad\qquad\qquad + 2\,[\text{kN}] \times 2\,[\text{m}]$
$\qquad\qquad\qquad\quad = 5R_A + 4 \,\text{kNm}$

anticlockwise moments $= 6[\text{kN}] \times 1[\text{m}] + 10[\text{kN}] \times 3[\text{m}]$
$$= 6 + 30 \text{ kNm}$$
$$= 36 \text{ kNm}$$

Equating clockwise and anticlockwise moments:
$$4 + 5R_A = 36$$
$$\therefore \quad 5R_A = 32$$
$$\therefore \quad R_A = \frac{32}{5} \frac{[\text{kNm}]}{[\text{m}]}$$
$$R_A = 6.4 \text{ kN}$$

As a check consider the vertical equilibrium of the beam.

upward forces $\quad = R_A + R_D = 6.4 + 11.6 = \quad 18 \text{ kN}$
downward forces $\quad = 10 + 6 + 2 \qquad\qquad = \quad 18 \text{ kN}$
Therefore, upward forces$=$ downward forces.

Thus, the answers for R_A and R_D are correct.

Self-assessment questions

33   Figure 244 shows a horizontal beam A, B, C, D, 7 m long, simply supported at A and D and loaded as shown. Determine the magnitude of
the reactions.

34   Figure 245 shows a horizontal beam A, B, C, D, E, 10 m long, simply
supported at A and D and loaded as shown. Determine the magnitude
of the reactions.

35   Figure 246 shows a horizontal beam A, B, C, D, E, 8 m long, simply
supported at B and D and loaded as shown. Determine the magnitude
of the reactions.

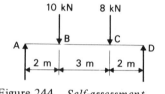

Figure 244   *Self-assessment
question 33*

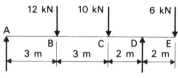

Figure 245   *Self-assessment
question 34*

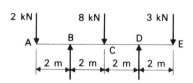

Figure 246   *Self-assessment
question 35*

Solutions to self-assessment questions

33   Let R_A and R_D be the reaction forces.
Take moments about an axis through A,

anticlockwise moments $= R_A[kN] \times 0[m] + R_D[kN] \times 7[m]$
$\qquad = 7R_D \text{ kNm}$

clockwise moments $\quad = 10[kN] \times 2[m] + 8[kN] \times 5[m]$
$\qquad = 20 + 40 \text{ kNm}$
$\qquad = 60 \text{ kNm}$

Equating clockwise and anticlockwise moments:
$$7R_D = 60$$
$$R_D = \frac{60}{7}\frac{[kNm]}{[m]} = 8.57 \text{ kN}$$

Take moments about an axis through D,

anticlockwise moments $= 8[kN] \times 2[m] + 10[kN] \times 5[m]$
$\qquad = 16 + 50 \text{ kNm}$
$\qquad = 66 \text{ kNm}$

clockwise moments $\quad = R_A[kN] \times 7[m] + R_D[kN] \times 0[m]$
$\qquad = 7R_A \text{ kNm}$

Equating moments,
clockwise moments $\quad =$ anticlockwise moments
$$7R_A = 66$$
$$R_A = \frac{66}{7}\frac{[kNm]}{[m]} = 9.43 \text{ kN}$$

Check by equating upward and downward forces,

upward forces $\quad = R_A + R_D = 9.43 + 8.57 = 18 \text{ kN}$
downward forces $\quad = 10 + 8 \qquad\qquad\qquad = 18 \text{ kN}$
therefore, upward forces $=$ downward forces.
The reaction forces are correct.

34   Let R_A and R_D be the reaction forces.
Take moments about an axis through A,

anticlockwise moments $= R_A[kN] \times 0[m] + R_D[kN] \times 8[m]$
$\qquad = 8R_D \text{ kNm}$

clockwise moments $\quad = 12[kN] \times 3[m] + 10[kN] \times 6[m]$
$\qquad\qquad + 6[kN] \times 10[m]$
$\qquad = 36 + 60 + 60 \text{ kNm}$
$\qquad = 156 \text{ kNm}$

Equating moments,
clockwise moments $\quad =$ anticlockwise moments
$$156 = 8R_D$$
$$\therefore \quad R_D = \frac{156}{8}\frac{[kNm]}{[m]} = 19.5 \text{ kN}$$

Take moments about an axis through D,

clockwise moments $\quad = R_A[kN] \times 8[m] + 6[kN] \times 2[m]$
$\qquad\qquad + R_D[kN] \times 0[m]$
$\qquad = 8R_A + 12 \text{ kNm}$

$$\text{anticlockwise moments} = 10[kN] \times 2[m] + 12[kN] \times 5[m]$$
$$= 20 + 60 \text{ kNm}$$
$$= 80 \text{ kNm}$$

Equating moments,

$$\text{clockwise moments} = \text{anticlockwise moments}$$
$$8R_A + 12 = 80$$
$$8R_A = 68$$
$$R_A = \frac{68}{8} \frac{[kNm]}{[m]} = 8.5 \text{ kN}$$

Check by equating upward and downward forces,

upward forces $= R_A + R_D = 8.5 + 19.5 \quad\quad = 28 \text{ kN}$

downward forces $= 12 + 10 + 6 \quad\quad\quad\quad = 28 \text{ kN}$

therefore, upward forces = downward forces.

The reaction forces are correct.

35 Let R_B and R_D be the reaction forces.

Take moments about an axis through B,

$$\text{anticlockwise moments} = R_B[kN] \times 0[m] + R_D[kN] \times 4[m]$$
$$+ 2[kN] \times 2[m]$$
$$= 4R_D + 4 \text{ kNm}$$

$$\text{clockwise moments} = 8[kN] \times 2[m] + 3[kN] \times 6[m]$$
$$= 16 + 18 \text{ kNm}$$
$$= 34 \text{ kNm}$$

equating moments,

$$\text{anticlockwise moments} = \text{clockwise moments}$$
$$4R_D + 4 = 34$$
$$4R_D = 30$$
$$R_D = \frac{30}{4} \frac{[kNm]}{[m]} = 7.5 \text{ kN}$$

Take moments about an axis through D,

$$\text{clockwise moments} = R_B[kN] \times 4[m] + 3[kN] \times 2[m]$$
$$+ R_D[kN] \times 0[m]$$
$$= 4R_B + 6 \text{ kNm}$$

$$\text{anticlockwise moments} = 8[kN] \times 2[m] + 2[kN] \times 6[m]$$
$$= 16 + 12 \text{ kNm}$$
$$= 28 \text{ kNm}$$

Equating moments,

$$\text{clockwise moments} = \text{anticlockwise moments}$$
$$4R_B + 6 = 28$$
$$4R_B = 22$$
$$R_B = \frac{22}{4} \frac{[kNm]}{[m]} = 5.5 \text{ kN}$$

Check by equating upward and downward forces,

upward forces $= R_B + R_D = 5.5 + 7.5 = 13 \text{ kN}$

downward forces $= 8 + 2 + 3 \quad\quad\quad = 13 \text{ kN}$

therefore, upward forces = downward forces.

The reaction forces are correct.

Section 4
Simple machines

After reading the following material, the reader shall:

4 Understand simple machines.

4.1 Describe a machine as a device for changing the magnitude and line of action of a force.

4.2 Define the terms:
  (*a*) force ratio (mechanical advantage),
  (*b*) movement ratio (velocity ratio),
  (*c*) efficiency.

4.3 Predict the units for force ratio and movement ratio.

4.4 Explain why the movement ratio of a machine is constant.

4.5 Deduce that the value of force ratio varies with the magnitude of the load.

4.6 Explain why the efficiency of a machine cannot reach 100%.

4.7 Deduce that the efficiency value varies with the magnitude of the load.

A machine is a device for changing the magnitude and line of action of a force. For example, when using a car jack, the circular movement of the hand-operated member is changed to the vertical movement of the car. A machine is also a device which enables work to be done more conveniently; for example, a properly designed lifting device enables large gravitational loads to be overcome by the application of quite small forces. Again, the car jack is an example of this property of a correctly designed lifting machine. Machines may be hand or power operated. Some examples of industrial lifting machines in common use are pulley blocks (which may be rope or chain operated), geared hoists and screw jacks. A bench vice is also an example of a simple machine.

Figure 247 shows a sketch and gives the leading dimensions of a lifting device, known as a worm and wheel, used in a college laboratory. A set of values for load and effort for this machine are given in rows 1 and 2 in the table below. Rows 3, 4 and 5 are to be completed as the reader works through the following pages.

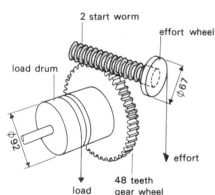

Figure 247 *Worm and wheel*

row number							
1	load N	0	10	20	30	40	50
2	effort N	2.8	5.5	8.2	10.9	13.6	16.3
3	force ratio						
4	movement ratio						
5	efficiency						

row number							
1	load N	60	70	80	90	100	110
2	effort N	19.0	21.7	24.4	27.1	29.8	32.5
3	force ratio						
4	movement ratio						
5	efficiency						

Self-assessment question

36   Complete the following statement and underline one word in each bracket which makes the statement correct:
From the values listed for load and effort in the table it is apparent that a relatively (LARGE/SMALL) effort will lift a (LARGE/SMALL) load.

By using a machine it is possible to gain an advantage in that a large load may be moved by a relatively small effort. The ratio of load to effort is known as the force ratio. That is: force ratio = load/effort.

Self-assessment questions

37   What are the units, if any, for force ratio?

38   What are the values of force ratio for each of the pairs of values for load and effort given in the table? Enter the answers in row 3 of the table.

39   Complete the following statement by underlining the word or words which make the statement correct:
From the set of values calculated for force ratio it can be concluded that as the load increases the force ratio (VARIES/REMAINS CONSTANT) for the machine.

Now consider the machine from a different viewpoint. How far does the load move for one revolution of the effort wheel? Referring to Figure 247 for the worm and wheel, let the effort wheel be turned through one complete revolution.

distance moved by effort   =   circumference of effort wheel
$$= \pi \times D$$
$$= \pi \times 67 \text{ mm}$$
$$= 210.5 \text{ mm}$$

One revolution of the worm causes each tooth on the gear wheel to be displaced along the circumference of the wheel a distance equal to the lead of the worm. Because the worm is a two-start thread, then the distance moved by the load is:

$$= \frac{2}{48} \times \text{ circumference of load drum}$$

$$= \frac{2}{48} \times \pi \times 92 \text{ mm}$$

$$= 12.04 \text{ mm}$$

Study the results for the distances moved by the load and the effort and then answer the following question.

Self-assessment question

40 Complete the following statement by underlining the word which makes the statement correct:
The distance moved by the effort is much (LARGER/SMALLER) than the distance moved by the load.

The ratio between the distance moved by the effort and the distance moved by the load is known as the movement ratio.

$$\text{movement ratio} = \frac{\text{distance moved by effort}}{\text{distance moved by load}}$$

Self-assessment question

41 What are the units, if any, for movement ratio?

As can be seen from the calculations for the distances moved by the effort and the load, these values depend only on the dimensions of the machine and therefore the movement ratio for any machine is a constant.

Self-assessment questions

42 Calculate the value of the movement ratio for the machine detailed in Figure 247.

Solutions to self-assessment questions

36 SMALL effort, LARGE load.

37 There are no units, the term is a ratio, e.g. newton/newton.

38

Row 3:						
force ratio	0	1.82	2.44	2.75	2.94	3.07
	3.16	3.23	3.28	3.32	3.36	3.38

39 The force ratio varies for the machine.

43 Complete the following statement by underlining the word or words that makes the statement correct:

Because the movement ratio for a particular machine depends upon the geometry of the machine, then it is true to say that the value of the movement ratio of a machine (VARIES/IS A CONSTANT).

It is impossible to get more work out of a machine than the work that is put into it. In fact the work output is always less than the work input because of the friction forces that are present between the contact surfaces. The difference between work input and work output represents wasted energy which should be kept to a minimum. The success achieved in this respect is a measure of the efficiency of the machine.

$$\text{efficiency} = \frac{\text{work output}}{\text{work input}}$$

Efficiency is usually expressed as a percentage in which case the ratio is multiplied by 100. The basic equation for work done is:

work done = force × distance moved by force in the direction of the force

So, for the machine:

work output = load × distance moved by load

and

work input = effort × distance moved by effort

Substituting the expressions for work output and work input into the equation for efficiency, then

$$\text{efficiency} = \frac{\text{load} \times \text{distance moved by load}}{\text{effort} \times \text{distance moved by effort}}$$

Now $\dfrac{\text{load}}{\text{effort}}$ = force ratio,

and $\dfrac{\text{distance moved by load}}{\text{distance moved by effort}} = \dfrac{1}{\text{movement ratio}}$

Therefore

$$\text{efficiency} = \frac{\text{force ratio}}{\text{movement ratio}}$$

Self-assessment question

44 What are the values of efficiency for the range of loads given in the table? Enter the answers in row 5.

Study the results obtained for the values of efficiency and then answer the following question.

Self-assessment question

45 Complete the following statement by underlining the word or words which makes the statement correct.
From the values obtained for the efficiency of the machine it is apparent that the value of the efficiency of a machine (VARIES/REMAINS CONSTANT) as the load increases.

Summary
The properties of a machine may be summarized as follows:

(a) a small effort lifts a relatively large load, this advantage being known as the force ratio.

$$\text{force ratio} = \frac{\text{load}}{\text{effort}}$$

(b) the force ratio varies with the load.

(c) the distance moved by the effort is much greater than the distance moved by the load. The ratio of these distances is known as the movement ratio.

$$\text{movement ratio} = \frac{\text{distance moved by effort}}{\text{distance moved by load}}$$

(d) the movement ratio is a constant for a machine and depends only upon the geometry of the machine.

(e) $$\text{efficiency} = \frac{\text{force ratio}}{\text{movement ratio}}$$

Solutions to self-assessment questions

40 The distance moved by effort is much LARGER than the distance moved by the load.

41 No units, the term is a ratio, e.g. metre/metre.

42 $$\text{Movement ratio} = \frac{\text{distance moved by effort}}{\text{distance moved by load}} = \frac{210.5}{12.04} = 17.48 \text{ metre/metre.}$$

43 The movement ratio of a machine is a constant. Now complete row 4 of the table by entering in the value of the movement ratio for the machine.

44 | Efficiency | 0 | 10.4 | 13.96 | 15.73 | 16.82 | 17.56 |
    | | 18.08 | 18.48 | 18.76 | 18.99 | 19.22 | 19.34 |

(*f*)   the efficiency of a machine varies with the load.

(*g*)   the value of the efficiency of a machine cannot reach 100% because of the friction forces present between the contact surfaces.

After reading the following material, the reader shall:

4.8   Use results obtained experimentally from a machine to plot graphs of effort, force ratio and efficiency against the load — the load line being the independent variable.

Self-assessment question

46   Using the results entered into the table on pages 166–7, plot the graphs of effort, force ratio and efficiency against the load, with the load being the independent variable as shown in Figures 248, 249 and 250.

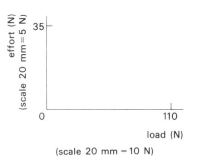

Figure 248   *Load-effort axes*

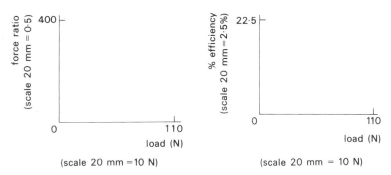

Figure 249   *Load-force ratio axes*   Figure 250   *Load-efficiency axes*

After reading the following material, the reader shall:

4.9   Use experimental results to determine the law of the machine in the form $E = aF + b$.

4.10   Explain that the intercept b is the effort required to move the machine when the load is zero.

4.11   State that limiting force ratio is $1/a$.

4.12   Deduce the limiting efficiency for a machine.

4.13   Identify the effort lost due to friction.

4.14   Predict the effort required to lift a load.

4.15   Predict the load that will be raised by a given effort.

4.16   Explain the term overhauling.

4.17   State that overhauling is not possible if the efficiency of the machine is less than 50%.

As discussed earlier, the movement ratio is fixed for any given machine and may be calculated from the geometry of the machine. The graphs shown in Figures 251, 252 and 253 show how effort, force ratio and efficiency vary with load.

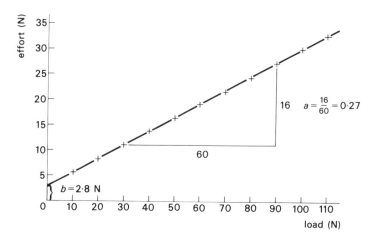

Figure 251   *Load-effort graph*

The basic characteristic of a machine is the variation of effort with load. For the machine detailed in Figure 247 the graph of load against effort is a straight line. This is true for all simple machines. Because the relationship between load and effort is a straight line, this relationship may be expressed in the form:

$$E = aF + b$$

where E = actual effort
        F = load
and a, b = constants

(this equation is just a variation on $y = mx + c$)

The equation $E = aF + b$ is known as the law of the machine.

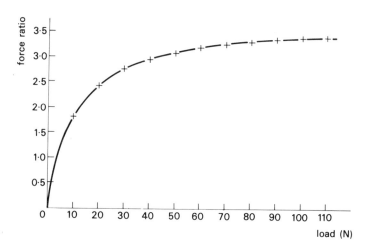

Figure 252   *Load-force ratio graph*

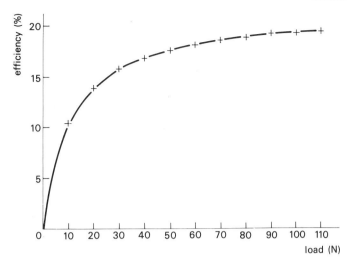

Figure 253 *Load-efficiency graph*

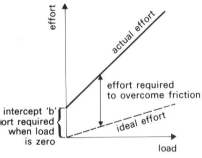

Figure 254 *Actual-ideal effort*

In an ideal machine the load-effort line passes through the origin as in Figure 254. In effect this means that the ideal machine has an efficiency of 100%, and in practice this means that the machine is frictionless. However, there are always friction forces present between the contact surfaces of the moving parts, and hence the actual load-effort line is displaced from the origin. The constant b is thus the effort required to overcome friction when there is no load on the machine. Note that the value of the friction forces increases as the load increases. This is because the normal reaction between the surfaces in contact increases.

Self-assessment questions

47 Using the graph of load-effort plotted on Figure 251 determine the law of the machine.

48 Using the load-effort graph read off the values of:
(*a*) the effort required to lift a load of 65 N,
(*b*) the load lifted by an effort of 17.5 N.

49 Check the answers read off the graph by calculation using the law of the machine.

The force ratio for a machine may be written as:

$$\text{force ratio} = \frac{\text{load}}{\text{effort}} = \frac{F}{aF + b}$$

Solutions to self-assessment questions

45 The value of the efficiency of a machine varies as the load increases.

46 See Figures 251, 252 and 253.

Divide top and bottom by F and the equation becomes:

$$\text{force ratio} = \frac{1}{a + b/F}$$

If the load F becomes very large, then the term b/F becomes negligible when compared with a. This means that the force ratio approaches a limiting value which is given by:

$$\text{force ratio (limiting)} = \frac{1}{a}$$

The equation gives the maximum value of force ratio that is theoretically possible for a given machine.

Self-assessment questions

50 Calculate the limiting value for the force ratio for the machine shown in Figure 247.

51 Mark on the load-force ratio graph shown in Figure 252 a horizontal line to represent the limiting value of the force ratio.

From page 169 the equation for efficiency is:

$$\text{efficiency} = \frac{\text{force ratio}}{\text{movement ratio}}$$

Now, because the movement ratio is a constant for a given machine, the variation of efficiency with load is similar to that of force ratio with load. This can be seen by comparing the shape of the graphs of efficiency and force ratio against load, drawn on Figures 252 and 253.

It has been shown that

$$\text{force ratio} = \frac{1}{a + b/F}$$

Solutions to self-assessment questions

47 From the graph $b = 2.8$.
Slope $a = 0.27$.
Therefore law of machine is $E = 0.27F + 2.8$

48 (a) 20.4 N
(b) 54 N

49 If $F = 65$ N,
then $E = 0.27 \times 65 + 2.8 = 20.35$ N
If $E = 17.5$ N,

then $F = \dfrac{17.5 - 2.8}{0.27} = 54.44$ N

If this is substituted for force ratio in the equation for efficiency, then the equation becomes:

$$\text{efficiency} = \frac{1}{\text{movement ratio}\,(a + b/F)}$$

Again, as F increases, then b/F decreases and may be neglected when compared with a for very large values of F. In the same way as a limiting value for force ratio was obtained, a limiting value for efficiency is given by

$$\text{efficiency (limiting)} = \frac{1}{\text{movement ratio} \times a}$$

This equation gives the maximum efficiency that is theoretically possible for a given machine.

Self-assessment questions

52 Calculate the value of the limiting efficiency for the machine shown in Figure 247.

53 Mark on the load-efficiency graph a horizontal straight line, to represent the limiting value of efficiency.

Whilst considering efficiency, it is appropriate to look at a feature of machines known as overhauling. If, when a load is being raised, the effort is removed from the input end of the machine and the load then begins to fall (thus reversing the original effect of the effort), the machine is said to overhaul.

It can be shown that if a machine is not to overhaul, then the efficiency of the machine must be less than 50%.

The reader will appreciate that if a machine has an efficiency of less than 50%, then over half the energy supplied is not put to useful work. This is not a very satisfactory situation and obviously manufacturers of industrial lifting machines try to achieve as high an efficiency value as possible for their machines. There remains then the problem of overhauling, which can be overcome in a number of ways. On cranes the hoist brakes are always on when there is no power being supplied to the motor. On a simple geared winch, overhauling can be prevented by the use of a ratchet arrangement.

Self-assessment questions

54 Write down the general expression for the law of a machine in terms of load F and effort E.

55 The term b in the equation for the law of a machine represents the effort required to move the machine when the magnitude of the load is

56   The equation for the limiting value of force ratio is given by the expression:

force ratio (limiting) = _____ .

57   The equation for the limiting value of efficiency is given by the expression:

efficiency (limiting) = _____ .

58   If the law of a lifting machine is given by the equation
$E = 0.2F + 0.15$ kN
   (*a*) Calculate the effort required to raise a load of 8 kN.
   (*b*) Calculate the load that can be lifted by an effort of 1.15 kN.
   (*c*) Calculate the value of the limiting force ratio for the machine.
   (*d*) Calculate the limiting value of efficiency for the machine if the movement ratio for the machine is 6.
   (*e*) State the value of the effort required to overcome the friction of the machine when the load is zero.

59   Explain what is meant by the term overhauling.

Solutions to self-assessment questions

50   Force ratio (limiting) $= \dfrac{1}{a} = \dfrac{1}{0.27} = 3.7$

51   The limiting value of the force ratio is illustrated on Figure 255. Note that the curve showing the actual force ratio values approaches this maximum value but does not reach it.

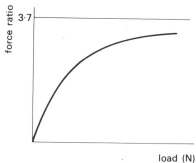

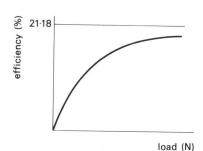

Figure 255   *Solution to self-assessment question 51*

Figure 256   *Solution to self-assessment question 53*

52   Efficiency (limiting) $= \dfrac{1}{17.48 \times 0.27} = 0.2118 = 21.18\%$

53   The limiting value of efficiency is illustrated on Figure 256. Note that the curve showing the actual efficiency values approaches this maximum value but does not reach it.

54 $E = aF + b$

55   Zero.

60　For a machine not to overhaul then the value of its efficiency must be less than＿＿＿＿＿＿＿ .

After reading the following material, the reader shall:

4.20　Describe with the aid of sketches the construction of:
(*a*) pulley system,
(*b*) screw jack,
(*c*) gear system.

4.21　Solve problems involving the quantities force ratio, movement ratio and efficiency in relation to simple machines.

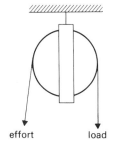

effort　　　　load

Figure 257　*Single pulley system*

Having used the results from a test on one machine to illustrate the various characteristics of machines, it is now necessary to broaden the discussion to consider some of the simple machines that are in use. It has been established that the movement ratio of a machine is a constant for a particular machine, and that the value of the movement ratio depends upon the geometry of the machine. The method of calculating the movement ratio is basically the same for all machines. The effort member is given some convenient displacement, say one complete revolution, and then, using the dimensions of the machine, the displacement of the load member is calculated. The movement ratio can then be calculated.

Some simple machines will now be analyzed using this basic approach to determine the movement ratio.

Pulleys
In a single-pulley system as shown in Figure 257, the rope attached to the load passes over the top of the pulley to where the effort is applied.

Self-assessment question

61　If the effort is moved through a distance of 2 m then how far is the load lifted? What is the movement ratio for this pulley system?

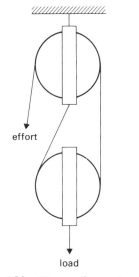

effort

load

Figure 258　*Two pulley system*

The arrangement for a two pulley system is shown in Figure 258; the rope passes from the fixed point round the bottom pulley and up and over the top pulley.

Self-assessment questions

62　If the effort is moved through a distance of x metres, how far is the load lifted?

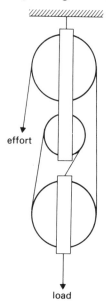

effort

load

Figure 259   *Three pulley system*

63   What is the movement ratio for this pulley system?

The arrangement for a three pulley system is shown in Figure 259.

Self-assessment question

64   What is the movement ratio for this pulley system?

It was shown that the movement ratio for a single-pulley system is 1, for a two-pulley system it is 2 and for a three-pulley system it is 3.

Self-assessment question

65   In general for a pulley system, what relationship is there between the total number of pulleys and the value of the movement ratio?

Example 20
A body having a mass of 60 kg is lifted by means of a pulley block arrangement as shown in Figure 260. If the upper and lower blocks have two pulleys and one pulley respectively, and the applied effort is 280 N, calculate:

Solutions to self-assessment questions

56    $1/a$

57    $\dfrac{1}{\text{movement ratio} \times a}$

58   (a) $E = 0.2 \times 8 + 0.15 = 1.75$ kN
(b) F
(b) $F = \dfrac{1.15 - 0.15}{0.2} = 5$ kN

(c) Limiting force ratio $= \dfrac{1}{a} = \dfrac{1}{0.2} = 5$ newton/newton

(d) Efficiency (limiting) $= \dfrac{1}{\text{movement ratio} \times a} = \dfrac{1}{6 \times 0.2}$

$= 0.833 = 83.3\%$
(e) 0.15 kN, i.e. the value of b from $E = aF + b$

59   If the effort is removed from a machine and the load begins to fall the machine is said to overhaul.

60   50%

61   The load is lifted through 2 metres; therefore the movement ratio
$= \dfrac{\text{distance moved by effort}}{\text{distance moved by load}} = \dfrac{2\text{ m}}{2\text{ m}} = 1$ metre/metre

62   The load moves a distance of $x/2$ metres, because both lengths of rope around the bottom pulley have to be shortened.

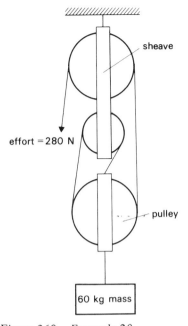

Figure 260   *Example 20*

(a) the movement ratio,
(b) the force ratio,
(c) the efficiency of the machine at this load.

(a) Movement ratio = $\dfrac{\text{distance moved by effort}}{\text{distance moved by load}}$

For a pulley arrangement:
    movement ratio = total number of pulleys
∴ movement ratio = 2 + 1
                    = 3

(b) Force ratio = $\dfrac{\text{load}}{\text{effort}}$

In this example, the load is the gravitational force mg.

$$\therefore \text{ load } = 60\,[\text{kg}] \times 9.81\left[\frac{\text{m}}{\text{s}^2}\right]$$
$$= 588.6\,[\text{kg}\frac{\text{m}}{\text{s}^2}]$$
$$= 588.6\,\text{N}$$

The effort required to lift this mass = 280 N

$$\therefore \text{ force ratio } = \frac{588.6\ [\text{N}]}{280\ [\text{N}]}$$
$$= 2.1\ \text{newton/newton}$$

(c) Efficiency = $\dfrac{\text{force ratio}}{\text{movement ratio}} = \dfrac{2.1}{3} = 0.7$

∴ efficiency = 0.7 or 70%

Screw jack

The screw jack shown in Figure 261 consists of a vertical screw which runs in a fixed nut. The fixed nut is part of the main body of the screw jack. The load is placed on the top and is caused to rise or fall when the effort lever arm is rotated.

The movement ratio = $\dfrac{\text{distance moved by effort}}{\text{distance moved by load}}$

Let R = radius of effort arm,
    p = pitch of the screw thread,
    n = number of starts.

Let the effort arm move through one complete revolution. Therefore, the distance moved by the effort = $2\pi R$. For one revolution of the effort arm the load will be lifted or lowered by an amount equal to the lead of the screw. That is, the distance moved by the load = np.

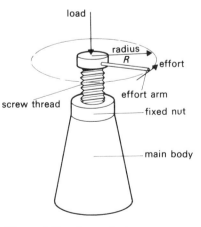

Figure 261   *Screw jack*

Therefore, the movement ratio $= \dfrac{2\pi R}{np}$

Example 21

A screw jack is used to support material that is being cut in a mechanical saw. If the screw jack has an effort arm of 175 mm radius and a single-start square thread of 6 mm pitch, calculate the efficiency of the jack if it requires a 10 N force to lift a mass of 40 kg.

$$\text{Efficiency} = \frac{\text{force ratio}}{\text{movement ratio}}$$

$$\text{force ratio} = \frac{\text{load}}{\text{effort}}$$

$$\text{The load} = mg = 40\,[\text{kg}] \times 9.81\,[\tfrac{m}{s^2}] = 392.4\,[\text{kg}\,\tfrac{m}{s^2}]$$

$$= 392.4\,\text{N}$$

This is lifted with an effort of 10 N

$$\therefore \quad \text{force ratio} = \frac{392.4\,[\text{N}]}{10\,[\text{N}]} = 39.24\ \text{newton/newton}$$

$$\text{movement ratio} = \frac{\text{distance moved by effort}}{\text{distance moved by load}}$$

For one revolution of the effort arm of radius 175 mm,

$$\text{distance moved by effort} = 2\pi \times 175\ \text{mm}$$
$$= 1\,099.6\ \text{mm}$$

Distance moved by load for one revolution of the effort arm is equal to the lead of the screw, which in this case is equal to the pitch of the thread because it is a single-start thread. Therefore, distance moved by load = 6 mm.

Solutions to self-assessment questions

63 movement ratio $= \dfrac{\text{distance moved by effort}}{\text{distance moved by load}}$

$$= \frac{x}{x/2} = 2\ \text{metre/metre}$$

64 Movement ratio = 3, because if the effort moves a distance of x metres, then the three ropes attached shorten by $x/3$ metres.

65 For a pulley system:
movement ratio = total number of pulleys.

$$\text{Therefore, movement ratio} = \frac{1\,099.6 \ \text{mm}}{6 \ \text{mm}}$$

$$= 183.3 \ \text{mm/mm}$$

$$\therefore \quad \text{efficiency} = \frac{\text{force ratio}}{\text{movement ratio}} = \frac{39.24}{183.3} = 0.214$$

$$\therefore \quad \text{efficiency} = 0.214 \ \text{or} \ 21.4\%$$

Self-assessment question

66 The table of a shaping machine is lifted by means of a single-start square thread screw with a 6 mm pitch. The effort is applied at the end of a handle 240 mm long. If the total mass of the table, vice and component is 325 kg, determine for an efficiency of 25%:

(i)   the movement ratio,
(ii) the force ratio,
(iii) the effort required to lift the 325 kg mass.

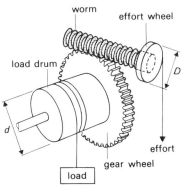

Figure 262   *Worm and wheel*

Worm and wheel

The arrangement of this type of machine is shown in Figure 262. The worm engages with the gear teeth on the wheel. One revolution of the worm causes each tooth on the wheel to be displaced along the circumference of the wheel a distance equal to the lead of the worm.

Let D = diameter of effort wheel,
d = diameter of load drum,
n = number of starts on worm,
p = pitch of teeth,
T = number of teeth on gear wheel.

In order to determine the expression for the movement ratio, consider one revolution of the effort wheel, which means one complete revolution of the worm.

Distance moved by effort $= \pi D$.

For one revolution of the worm, each tooth on the gear wheel is displaced along the circumference of the wheel by a distance equal to the lead of the worm $= np$.

Therefore, the corresponding revolutions of the gear wheel

$$= \frac{np}{\text{circumference of gear wheel}}$$

$$= \frac{np}{Tp} = \frac{n}{T}$$

Because the revolutions of the gear wheel are equal to the revolutions of the load drum,

displacement of load = revolutions of load drum ✕
circumference of load drum

$$= \frac{n}{T} \times \pi d$$

Therefore, movement ratio $= \dfrac{\text{distance moved by effort}}{\text{distance moved by load}}$

$$= \frac{\pi D}{n/T \times \pi d} = \frac{TD}{nd}$$

Example 22

A single start worm and worm wheel similar to that shown in Figure 262 has an effort wheel of 140 mm diameter and a load drum diameter of 125 mm. If the worm wheel has 40 teeth calculate:

(*a*) the movement ratio for the machine,

(*b*) the effort required to raise a load of 400 kg if the efficiency of the machine is 67%.

(*a*) Movement ratio $= \dfrac{\text{distance moved by effort}}{\text{distance moved by load}}$

Let the effort wheel rotate one revolution.

Distance moved by effort $= \pi D$ where D = diameter of effort wheel.

Distance moved by load $= \dfrac{n}{T}\pi d$

where d = diameter of load drum,

n = number of starts on worm,

T = number of teeth on gear wheel.

Therefore movement ratio $= \dfrac{\pi D}{n/T \times \pi d} = \dfrac{TD}{nd}$

$$= \frac{40 \times 140 \ [\text{mm}]}{1 \times 125 \ [\text{mm}]}$$

$$= 44.8 \ \text{mm/mm}$$

(*b*) Efficiency $= \dfrac{\text{force ratio}}{\text{movement ratio}}$

Solution to self-assessment question

66 (i) Movement ratio $= \dfrac{2\pi \times 240}{6} = 251.3 \ \text{mm/mm}$

(ii) Force ratio = efficiency ✕ movement ratio
= 0.25 ✕ 251.3 = 62.83 newton/newton

(iii) effort $= \dfrac{\text{load}}{\text{force ratio}} = \dfrac{325 \times 9.81}{62.83} = 50.74 \ \text{N}$

Therefore, force ratio   = efficiency × movement ratio

= 0.67 × 44.8

= 30

Now, force ratio $= \dfrac{\text{load}}{\text{effort}}$

Rearranging to give: effort $= \dfrac{\text{load}}{\text{force ratio}}$

In this case, load $= 400\,[\text{kg}] \times 9.81\,[\dfrac{\text{m}}{\text{s}^2}]$

$= 3\,924\,[\text{kg}\,\dfrac{\text{m}}{\text{s}^2}]$

Therefore, effort required $= \dfrac{\text{load}}{\text{force ratio}}$

$= \dfrac{3\,924}{30}\,\text{N}$

$= 130.8\,\text{N}$

Self-assessment question

67   A worm and wheel lifting device has a two-start worm and a wheel with 60 teeth. The load drum has a diameter of 0.25 m. Determine the force required at the 0.5 m diameter effort wheel to raise a mass of 250 kg if the efficiency of the machine is 40%.

Wheel and axle

Figure 263 shows a wheel and axle arrangement. The effort wheel of diameter D and the axle of diameter d are carried on the same shaft. The effort is applied to a cord wound round the rim of the effort wheel. To determine an expression for the movement ratio of the wheel and axle, consider one complete revolution of the effort wheel.

Distance moved by effort $= \pi D$

Because the wheel and axle are on the same shaft, then if the wheel makes one complete revolution, the axle must do the same.

Therefore, distance moved by load $= \pi d$

movement ratio $= \dfrac{\text{distance moved by effort}}{\text{distance moved by load}}$

$= \dfrac{\pi D}{\pi d} = \dfrac{D}{d}$

movement ratio $= \dfrac{D}{d}$

An extension of the wheel and axle is the differential wheel and axle.

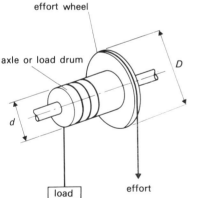

effort wheel

axle or load drum

D

d

load   effort

Figure 263   *Wheel and axle*

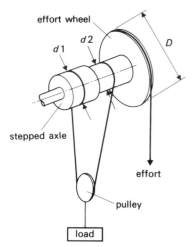

Figure 264   *Differential wheel and axle*

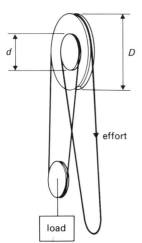

Figure 265   *Weston differential pulley block*

Differential wheel and axle

Figure 264 shows a differential wheel and axle. The axle is stepped, and each end of a cord is attached to and wrapped around the two different sections of the axle. The load is carried by a pulley at the bottom of the loop.

Consider one complete revolution of the effort wheel; the axle will also rotate one complete revolution. The looped cord unwinds from the smaller axle of diameter d_2, whilst the other end winds onto the larger axle of diameter d_1. For one complete revolution the loop shortens by an amount

$$\pi d_1 - \pi d_2 = \pi(d_1 - d_2)$$

This reduction in the loop is shared by the cord on either side of the pulley, so that the load is caused to rise by half this amount.

That is, distance moved by load $= \dfrac{\pi}{2}(d_1 - d_2)$

Therefore, movement ratio $= \dfrac{\text{distance moved by effort}}{\text{distance moved by load}}$

$$\text{movement ratio} = \frac{\pi D}{\pi/2(d_1 - d_2)} = \frac{2D}{(d_1 - d_2)}$$

The principle of the differential wheel and axle is used in practice with the Weston differential pulley block as shown in Figure 265.

In the Weston differential pulley block, an endless cord is wrapped round a two-diameter pulley in the top sheave, and carries in its lower loop the single bottom pulley. As the top two-diameter pulley rotates it tends to shorten the hanging loop, but the loop is also lengthened as the cord unwinds from the smaller diameter pulley. As with the differential wheel and axle, this leads to a total shortening of the loop which is shared by the cord on either side of the bottom pulley, so that the load is raised by an amount equal to half the total shortening of the cord.

Thus, the movement ratio $= \dfrac{\text{distance moved by effort}}{\text{distance moved by load}}$

$$\text{movement ratio} = \frac{\pi D}{\pi/2\,(D - d)} = \frac{2D}{(D - d)}$$

Solution to self-assessment question

67   Movement ratio = 60 (remember a two-start thread)
　　　Force ratio　　= efficiency × movement ratio = 24

　　　Effort　　　$= \dfrac{\text{load}}{\text{force ratio}} = \dfrac{250 \times 9.81}{24} = 102.2 \text{ N}$

Weston differential chain blocks usually have their rims shaped to suit the shape of the chain links, so instead of the pulley rim being curved, it is made up of a series of flats. The numbers of flats on each pulley can be used instead of the pulley diameters to calculate the movement ratio.

Example 23

A set of Weston differential pulley blocks has one pulley of 150 mm diameter and the other of 200 mm diameter. If the efficiency of the pulley blocks is 50% calculate:

(*a*) the movement ratio,

(*b*) the effort required to lift a component of mass 500 kg.

(*a*) Consider one revolution of the two pulley system shown in Figure 265.

Distance moved by effort $= \pi D$

Distance moved by load $=$ half the amount that the load support cord is shortened

$$= \frac{1}{2}(\pi D - \pi d) = \frac{\pi}{2}(D - d)$$

Therefore, movement ratio $= \dfrac{\text{distance moved by effort}}{\text{distance moved by load}}$

$$= \frac{\pi D}{\pi/2\,(D - d)} = \frac{2D}{(D - d)}$$

movement ratio $= \dfrac{2 \times 200 \quad [\text{mm}]}{(200 - 150)\,[\text{mm}]}$

$$= \frac{400}{50}$$

$$= 8 \text{ mm/mm}$$

(*b*) Efficiency $= \dfrac{\text{force ratio}}{\text{movement ratio}}$

Rearranging this equation so that:

force ratio $=$ efficiency $\times$ movement ratio

force ratio $= 0.5 \times 8$

$= 4$

Now, force ratio $= \dfrac{\text{load}}{\text{effort}}$

Rearranging this equation so that:

effort $= \dfrac{\text{load}}{\text{force ratio}}$

$$\text{effort} = \frac{500\,[\text{kg}] \times 9.81\,[\text{m/s}^2]}{4}$$

$$= 1226\ [\text{kg}\,\frac{\text{m}}{\text{s}^2}]$$

$$= 1\,226\,\text{N} = 1.226\,\text{kN}$$

Self-assessment question

68   The differential pulleys of a Weston chain block have 18 and 15 chain flats respectively, and the efficiency of the blocks is 72%. Find the force ratio of the blocks, and calculate the effort required to lift a fan shaft of mass 120 kg from its bearings.

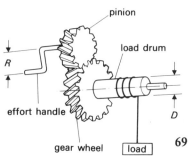

pinion

load drum

R

effort handle

D

gear wheel   load

Figure 266   *Geared winch*

Geared winch

The winch is a common mechanism used for winding and hauling; it may be powered or hand-operated. A hand-operated winch is shown in Figure 266. The effort is applied at the end of the effort arm, and cable is wound onto the drum. The simple or compound gear train is used to increase the movement ratio and so increase the mechanical advantage.

Self-assessment question

69   For the geared winch shown in Figure 266, let

R = radius of the effort arm,
D = diameter of the load drum,
T_1 = number of teeth on gear wheel,
T_2 = number of teeth on pinion.

Derive an expression for the movement ratio of the machine.

Example 24

For the winch shown in Figure 266, the radius of the effort arm is 350 mm and the diameter of the load drum is 125 mm. The pinion has 12 teeth and the gear wheel has 60 teeth. Calculate:

(a) the movement ratio for the winch,
(b) the efficiency of the winch if a load of 300 kg is lifted by a force of 150 N.

(a) Consider one revolution of the effort arm,

distance moved by effort   = $2\pi R$

= $2\pi \times 350$ mm

= $2\,199$ mm

For one revolution of the effort arm, the drum rotates   = $\frac{12}{60} \times 1$

= $\frac{1}{5}$ revolution

Therefore, distance moved
by load
$$= \frac{1}{5} \times \pi \times 125 \text{ mm}$$

$$= 78.54 \text{ mm}$$

$$\text{movement ratio} = \frac{\text{distance moved by effort}}{\text{distance moved by load}}$$

$$= \frac{2\,199 \text{ mm}}{78.54 \text{ mm}}$$

$$= 28 \text{ mm/mm}$$

(*b*) Efficiency $= \dfrac{\text{force ratio}}{\text{movement ratio}}$

Now, force ratio $= \dfrac{\text{load}}{\text{effort}}$

$$= \frac{300\,[\text{kg}] \times 9.81\,[\text{m/s}^2]}{150\,[\text{N}]}$$

$$= 19.62 \frac{[\text{kg} \times \text{m/s}^2]}{[\text{N}]}$$

$$= 19.62 \text{ newton/newton}$$

Therefore efficiency $= \dfrac{19.62}{28}$

$$= 0.7 \text{ or } 70\%$$

Self-assessment question

70   A simple hand-operated winch similar to that shown in Figure 266 has an effort arm of length 0.25 m and a load drum diameter of 100 mm. The number of teeth on the pinion is 50 and on the gear wheel 250. Calculate for an efficiency of 20%:
(*a*) the movement ratio,
(*b*) the force ratio,
(*c*) the load that can be raised by an effort of 500 N.

Further self-assessment questions

Stress and strain

71   An overhead crane is to have four ropes of equal diameter on which to support the load. The maximum load to be carried is 20 tonnes. If the ropes are to be made from steel in which the maximum permissible stress is limited to 100 N/mm², calculate the minimum suitable rope diameter.

72   A rod 50 mm diameter and 2 m long is subjected to tensile forces. The rod material has a modulus of elasticity of 200 kN/mm². The maximum stress in the rod material is not to exceed 100 N/mm² which is below the limit of proportionality of the material. Calculate the maximum forces that the rod can carry and the extension of the rod under the action of these forces.

73   A tubular tie bar is subjected to forces of 30 kN. The outer diameter of the bar is 30 mm and the maximum stress in the material is to be limited to 64 N/mm², which is below the limit of proportionality of the material. Calculate the inside diameter of the tubular tie bar and the extension of a 300 mm length of the bar. The value of the modulus of elasticity for the bar material is 85 kN/mm².

74   Calculate the forces required to punch a hole 75 mm diameter in a steel sheet 2 mm thick, if the shear strength of the material is 400 N/mm². What is the compressive stress in the 70 mm diameter shank of the punch due to these forces?

75   A hollow metal tube of square cross-section, 0.2 m long has its internal sides measuring 150 mm and is to carry forces of 1 MN. If the safe working stress is to be half the stress at the limit of proportionality which is 240 N/mm², calculate:

Solutions to self-assessment questions

68   Movement ratio = 12
      force ratio   = efficiency × movement ratio   = 0.72 × 12
                                                       = 8.64
      effort = $\dfrac{\text{load}}{\text{force ratio}}$ = $\dfrac{120 \times 9.81}{8.64}$ = 136.25 N

69   For one revolution of the effort handle,
      distance moved by effort   = $2\pi R$

      revolutions of load drum   = $\dfrac{T_2}{T_1}$

      distance moved by load   = $\dfrac{T_2}{T_1}\pi D$

      ∴   movement ratio   = $\dfrac{2\pi R}{T_2/T_1\,\pi D}$

                             = $\dfrac{2RT_1}{DT_2}$

70 (*a*) Movement ratio = $\dfrac{2 \times R \times T_1}{D \times T_2}$ = $\dfrac{2 \times 250 \times 250}{100 \times 50}$ = 25 mm/mm

    (*b*) Force ratio = efficiency × movement ratio
                        = 0.2 × 25 = 5
    (*c*) Load = effort × force ratio = 500 × 5 = 2.5 kN

71   Total area = 1 962 mm², area of one rope = 490.5 mm², diameter = 25 mm.

(*a*) the area of the metal required to carry the 1 MN force,

(*b*) the thickness of the walls of the column.

76 A metal tube of outside diameter 80 mm and length 2 m is subjected to tensile forces of 70 kN. If the safe working stress is 75 N/mm^2, which is less than the stress at the limit of proportionality, calculate the inside diameter of the tube. If the modulus of elasticity for the material is 100 kN/mm^2, calculate the amount the tube will extend due to tensile forces of 70 kN.

77 A crane is designed to carry a maximum mass of 90 tonnes and has eight ropes, made of material which has an ultimate strength of 600 N/mm^2, on which to support the mass. Using a working stress equal to one sixth the ultimate strength, calculate a suitable rope diameter.

78 Holes 25 mm square are to be punched in 8 mm thick plates made from material having a maximum shear strength of 300 N/mm^2. Calculate the maximum compressive stress in the 50 mm diameter punch shank.

79 A 300 mm hacksaw blade, 12 mm wide and 0.8 mm thick, is subjected to forces of 900 N when tightened in its frame on pegs 3.5 mm diameter. Calculate:

(*a*) the maximum tensile stress in the blade,

(*b*) the shear stress in the pegs.

Polygon of forces and simply supported beams

80 Figure 267 shows a system of forces acting at a point. Find the magnitude and direction of the resultant force.

81 Find the magnitude and direction of the resultant force due to the system of forces shown in Figure 268.

82 A mass of 1 tonne is suspended by two cords as shown in Figure 269. Determine the forces in the two cords.

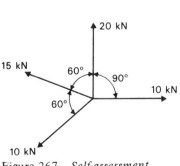

Figure 267 *Self-assessment question 80*

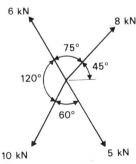

Figure 268 *Self-assessment question 81*

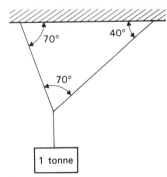

Figure 269 *Self-assessment question 82*

83   Find the magnitude and direction of the equilibrant force for the system of forces shown in Figure 270.

84   Find the magnitude and direction of the equilibrant force for the system of forces shown in Figure 271.

85   For each of the loaded beams shown in Figure 272 calculate the value of the reaction forces.

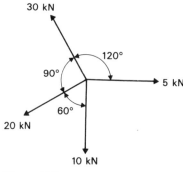

Figure 270   *Self-assessment question 83*

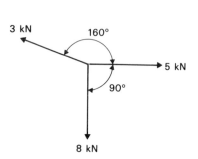

Figure 271   *Self-assessment question 84*

86   A horizontal beam ABCDEF is simply supported at B and E. AB = 2 m; BC = 5 m; CD = 3 m; DE = 2 m; EF = 2 m. Vertical forces of 5 kN, 10 kN, 8 kN and 3 kN are applied at A, C, D and F respectively. Calculate the magnitude of the reaction forces at B and E.

Solutions to self-assessment questions

72   Forces = 196.34 kN, extension = 1 mm

73   Total area = 468.75 mm², inside diameter = 20.77 mm, extension = 0.226 mm.

74   Stress = 48.98 N/mm².

75   (a) 8 333.33 mm², (b) 12.8 mm.

76   Inside diameter = 72.2 mm, extension = 1.5 mm.

77   Rope diameter = 37.48 mm.

78   Stress = 122.28 N/mm².

79   (a) Stress = 132.25 N/mm² – remember that the maximum stress occurs in the section of the blade where the hole is positioned.
      (b) Shear stress = 93.54 N/mm².

80   25.4 kN at 117.5° to the horizontal.

81   2.15 kN at 274.5° to the horizontal.

82   8 kN, 3.57 kN.

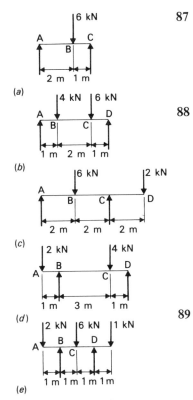

(a)

(b)

(c)

(d)

(e)

Figure 272 *Self-assessment question 85*

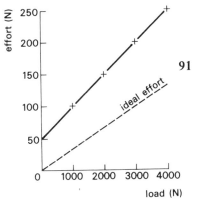

Figure 273 *Self-assessment question 88*

87 A horizontal beam AB is 15 m long and is simply supported at A and B. Forces of 6 kN, 10 kN and 8 kN are placed on the beam at distance 3 m, 6 m and x metres from A. Calculate the distance x if the two reaction forces are the same.

Simple machines

88 The graph in Figure 273 shows the relationship between force and effort for a machine, together with the ideal effort curve. Determine for the machine:

(a) the law of machine,

(b) the force required to overcome friction in the machine when there is zero load,

(c) the effort required to lift a load of 254.8 kg,

(d) the load that can be lifted by an effort of 225 N,

(e) the force ratio when the load is 203.8 kg,

(f) the limiting force ratio,

(g) the limiting efficiency if the movement ratio is 30,

(h) the effort lost due to friction when the load is 152.9 kg.

89 An electric motor of mass 100 kg is to be lifted from a bench by means of rope blocks. There are two pulleys in the upper block and three in the lower block. If the force ratio at this load is 4.5, calculate:

(a) the effort required to lift the motor,

(b) the movement ratio of the rope blocks,

(c) the efficiency of the rope blocks at this load.

90 A differential wheel and axle, similar to that shown in Figure 264, has an effort wheel diameter of 200 mm, and the stepped axle has diameters of 70 mm and 60 mm. Calculate:

(a) the movement ratio of the machine,

(b) the load that can be lifted by an effort of 120 N, if the force ratio is 24,

(c) the efficiency of the machine for this effort.

91 The following table gives values of load and effort obtained during an experiment on a machine:

load (kg)	150	300	450	600	750	900
effort (N)	185	243	310	360	413	480

The movement ratio of the machine is 40. Determine for the machine:

(a) the law of the machine,

(b) the limiting value of force ratio,

(c) the limiting efficiency value.

Solutions to self-assessment questions

83 18.67 kN at 198.7° to the horizontal.

84 7.3 kN at 87.4° to the horizontal.

85 (a) R_A = 2 kN R_C = 4 kN
    (b) R_A = 4.5 kN R_D = 5.5 kN
    (c) R_A = 2 kN R_C = 6 kN
    (d) R_B = 3.5 kN R_D = 2.5 kN
    (e) R_B = 5.5 kN R_D = 3.5 kN

86 R_B = 12 kN, R_E = 14 kN

87 12.75 m.

88 (a) Law of the machine, $E = aF + b$.
       b = intercept = 50, a = slope = 0.05, $E = 0.05F + 50$
    (b) force required = effort when load is zero = 50 N
    (c) from graph, effort = 175 N
    (d) from graph, effort of 225 N lifts a load of 3 500 N

    (e) force ratio = $\dfrac{\text{load}}{\text{effort}} = \dfrac{2\,000}{150}$ = 13.33 newton/newton

    (f) limiting force ratio = $\dfrac{1}{a} = \dfrac{1}{0.05}$ = 20

    (g) limiting efficiency = $\dfrac{1}{\text{movement ratio} \times a}$

                 = $\dfrac{1}{30 \times 0.05}$ = 0.667

    (h) Effort lost due to friction is the difference between the actual and ideal
       effort values; at 1.5 kN this is 125 − 50 = 75 N.

89 (a) Effort = $\dfrac{\text{load}}{\text{force ratio}} = \dfrac{100 \times 9.81}{4.5}$ = 218 N

    (b) Movement ratio = number of pulleys = 5
    (c) Efficiency = 0.9 or 90%.

90 (a) Movement ratio = $\dfrac{2D}{d_1 - d_2}$ = 40 mm/mm

    (b) Load = 2 880 N
    (c) Efficiency = 0.6 or 60 %.

91 (a) $E = 0.041F + 122$

    (b) limiting force ratio = $\dfrac{1}{a} = \dfrac{1}{0.041}$ = 24.39

    (c) limiting efficiency = $\dfrac{1}{40 \times 0.041}$ = 0.6097

Topic area: Dynamics

Section 1   Relative velocity

After reading the following material, the reader shall:

1   Know that velocities can be added vectorially.
1.1   State that velocity is a vector.
1.2   Solve simple problems involving relative velocities.

A *scalar* quantity possesses magnitude only; for example, speed, mass, energy, temperature and efficiency are scalar quantities. A *vector* quan- possesses both magnitude and direction; examples include velocity, acceleration and force.

The difference between speed (a scalar quantity) and velocity (a vector quantity) is that speed refers only to magnitude (e.g. 45 km/hour), whereas velocity refers to magnitude and direction (e.g. 45 km/hour in, say, the northerly direction). Note that the velocity of a vehicle travelling at 45 km/hour due north changes if either the magnitude or the direction changes.

The distinction between speed and velocity is not too difficult to perceive when examining linear motion, but it is perhaps less clear when considering angular motion. The speed of rotation of a crankshaft is defined by quoting a magnitude, such as 2 000 rev/min; the velocity of rotation of a crankshaft is defined by quoting both the magnitude and direction — 2 000 rev/min clockwise, viewed from the flywheel.

The motion of an aircraft may be specified as 1 000 km/hour towards New York. Does this statement describe the velocity of the aircraft? The answer is 'No', since the aircraft can be travelling in *any* direction towards New York. In order to describe the velocity of the aircraft, a reference point must also be specified, e.g. 1 000 km/hour towards New York from, or *relative to*, London. The phrase *relative to* is the key to this description; all velocities are relative to a chosen point or datum. In the above example, the datum is London.

The datum need not be a stationary point on the surface of the earth. Consider two aircraft: aircraft A is travelling at 1 000 km/hour due west, while aircraft B is travelling at 1 000 km/hour due east, but hopefully not on a collision course with aircraft A! The velocity of aircraft A relative to aircraft B is (1 000 + 1 000) km/hour due west, i.e. 2 000

km/hour due west. Expressed in symbols, this statement becomes:
velocity of A relative to B is $v_{AB} = 2\,000$ km/hour $\leftarrow$

Similarly, the velocity of aircraft B relative to aircraft A is 2 000 km/hour due east, or:
velocity of B relative to A is $v_{BA} = 2\,000$ km/hour $\rightarrow$

Another way of expressing the same relationship is to say that the velocity of B relative to A is the velocity with which B appears to be moving as seen by an observer situated on A and moving with A.

As velocity is a vector quantity, a vector diagram can be drawn to represent relative velocities. The vector diagram for the two aircraft discussed above is constructed as follows:

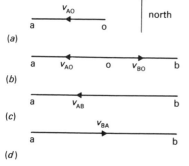

(a)

(b)

(c)

(d)

Figure 274   *Vector diagram of relative velocities*

A fixed point o is chosen (see Figure 274 (*a*)), and the vector $\overleftarrow{ao}$ is drawn from point o to represent the velocity of aircraft A, relative to a stationary point, in magnitude and direction. Remember that the vector is drawn from o to a, i.e. it is drawn from the datum. In a similar manner, the vector $\overrightarrow{ob}$ (Figure 274 (*b*)) is drawn to represent the velocity of aircraft B, relative to a stationary point, in magnitude and direction. The magnitude of the velocity of A relative to B (or B relative to A) is represented by the length ab. The direction of the velocity can be found in the following manner:

Velocity of A relative to B
The datum, or fixed point is b on the vector diagram; a vector is drawn from b to a (see Figure 274 (*c*)), and an arrow is added in the same direction (i.e. from b to a).

Velocity of B relative to A
The datum is now a on the vector diagram. Hence, the vector is drawn from a to b (see Figure 274 (*d*)), and an arrow added in the same direction.

In the above example, the relative velocities of two bodies travelling in opposite directions have been examined. Suppose that two cars are travelling down a straight road in the same direction, the speed of car X being 80 km/hour and the speed of car Y being 70 km/hour. The magnitude of the relative velocity of X relative to Y (or Y relative to X) is 10 km/hour, as can be seen from Figure 275 (*a*). Figure 275 (*b*) illustrates the velocity of X relative to Y (v_{XY}). Since y is the datum, the velocity v_{XY} is in the direction y $\rightarrow$ x. The velocity of Y relative to X (v_{YX}) is illustrated in Figure 275 (*c*), which shows the direction of v_{YX} in the direction x $\rightarrow$ y.

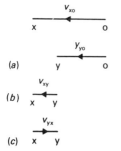

(a)

(b)

(c)

Figùre 275   *Relative velocities*

Example 1
The saddle of a lathe is moving towards the headstock at a speed of 0.5 mm/s. At the same time, the cross-slide is moving at 90° to the

direction of motion of the saddle at a speed of 0.15 mm/s. Determine the velocity of the cross-slide relative to the bed of the lathe.

Let A represent the saddle, B the cross-slide and O the fixed bed of the lathe.

$$\therefore \quad v_{AO} = 0.5 \text{ mm/s}$$
$$v_{BA} = 0.15 \text{ mm/s}$$

The vector diagram (Figure 276) is constructed by drawing $\overleftarrow{ao}$ to represent the velocity of the saddle relative to the bed, and $\overleftarrow{ba}$ to represent the velocity of the cross-slide relative to the saddle. Note that the second vector is not drawn from o, but from a, since it represents the velocity of B relative to A.

The vector $\overrightarrow{ob}$ represents the velocity of the cross-slide relative to the bed of the lathe; scaling from Figure 276, it is of magnitude 0.525 mm/s, in a direction 17° to the direction of motion of the saddle.

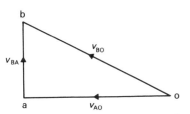

Figure 276 *Relative velocity of saddle and cross-slide*

Example 2

Two ships leave a harbour at the same time; ship A travels south east at a speed of 20 km/hour; ship B travels east at a speed of 25 km/hour. Find the velocity at which ship A appears to be travelling as seen by an observer on ship B.

Figure 277 shows the vector diagram for the velocities of the two ships; $\overrightarrow{ob}$ represents the velocity of B relative to a fixed point, and $\overrightarrow{oa}$ represents the velocity of A relative to a fixed point.

The velocity of A relative to B (i.e. the velocity at which ship A appears to be travelling as seen by an observer on ship B), is represented by the vector $\overrightarrow{ba}$, in the direction b → a. Scaling from Figure 277, $\overrightarrow{ba}$ is equivalent to a speed of 17.8 km/hour.

Thus, $v_{AB} = 17.8$ km/hour in a direction 7½° west of south.

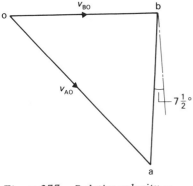

Figure 277 *Relative velocity – navigation*

In the above example, the calculation of relative velocity is not simply something of passing interest to observers on either ship. If the circumstances are modified slightly so that the two ships are on courses that are converging, rather than on courses which are diverging, then the accurate (and speedy!) derivation of relative velocity is of considerable interest to the navigator of each ship.

Vector diagrams involving relative velocities can also be used to simplify the complex analysis of mechanisms, such as the forces involved on the crankshaft and connecting rods of the internal combustion engine, or the behaviour of turbine blades. Another application of relative velocity occurs during the calculations necessary to rendezvous a rocket from earth with an orbiting space laboratory. This operation involves considerably more complex calculations than those considered here, but the basic principles are the same.

Self-assessment questions

1   Most countries impose 'speed limits' upon motor vehicles, a typical value being approximately 50 km/hour. Is this strictly a speed limit or a velocity limit?

2   Select from the following list those quantities which are vector quantities: mass, efficiency, velocity, pressure, speed.

3   Two motor vehicles are travelling in opposite directions along a straight stretch of motorway. Car A is travelling at 120 km/hour due south, and car B is travelling at 100 km/hour due north. Select from the following four alternatives the statement which describes the velocity of car B, relative to car A:
    (i)   20 km/hour due north,
    (ii)   220 km/hour due north,
    (iii) 220 km/hour due south,
    (iv)   20 km/hour due south.

4   The courses of two ships are plotted from a radio beacon, the angle between the courses being 30°. If ship A is travelling at 15 km/hour due north, and ship B at 18 km/hour, in a direction 30° east of north, find the velocity of ship B as it would appear to an observer on ship A.

5   Two aircraft are heading for London Airport. Aircraft A is travelling at 800 km/hour due east; aircraft B is travelling at 700 km/hour due north. Find the velocity of aircraft A relative to aircraft B. In what respect does the velocity of aircraft B relative to aircraft A differ from the answer?

6   An athlete has a rather unusual element in his training programme. He runs up an escalator, or moving staircase, which is moving down. The speed of the escalator relative to a fixed point is 0.5 m/s, and its true length measured along the incline is 24 m; the average velocity (relative to a fixed point) of the athlete running up the escalator is 8.5 m/s. Find the time for the athlete to run from the bottom to the top of the escalator.

7   A large casting is being transported across a workshop, by means of an overhead crane. If the crane is travelling across the workshop at 0.25 m/s, and the casting is being lowered at a rate of 0.2 m/s, determine the velocity of the casting relative to the floor of the workshop.

8   The pilot of a helicopter travelling at 100 km/hour in the north direction observes another helicopter which appears to be travelling at 120 km/hour in a south easterly direction. Determine the true velocity of the second helicopter.

Section 2
Angular motion

After reading the following material, the reader shall:

2   Solve problems for angular motion excluding variable acceleration.
2.1 Identify the radian as a unit of angular measurement.
2.2 Define angular velocity.
2.3 State the relation between linear and angular velocity.
2.4 State the relation between linear and angular acceleration.
2.5 Solve simple problems on angular velocity and acceleration.

The reader may well be familiar with a unit of angular speed or velocity such as revolutions per minute (rev/min). If a shaft rotates at a constant speed of 3 000 rev/min, it means that the shaft rotates 3 000 complete revolutions in one minute. Unfortunately, although the number of revolutions in a given time is relatively easy to measure, rev/min or rev/s do not form part of a coherent set of units. Instead of revolutions, it is conventional to express the speed or velocity of rotation in terms of the angle turned through in one second.

The radian
A unit of angular displacement which is consistent with units such as S.I. is the *radian*. The radian is the angle subtended at the centre of a circle by an arc equal to the radius. This definition is illustrated in Figure 278, where the angle XOY is equal to 1 radian.

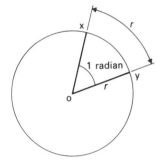

A circle of radius r has a circumference of length $2\pi r$. Therefore, there are 2π radians contained in one revolution, i.e.
1 revolution = 2π radians

Also, 2π radians = $360°$.
It follows that if the speed of a shaft is quoted as 50 rev/s, then its speed in radians per second (rad/s) is $50 \times 2\pi$ rad/s.

Figure 278   *Definition of the radian*

Angular velocity
Linear velocity is defined as the rate of change of linear displacement with respect to time. In certain circumstances, the definition can be simplified to:

$$\text{linear velocity} = \frac{\text{change in linear displacement}}{\text{time taken for the change}}$$

In a similar manner, the definition of angular velocity is *the rate of change of angular displacement with respect to time*, which can sometimes be written as:

$$\text{angular velocity} = \frac{\text{change in angular displacement}}{\text{time taken for the change}}$$

If ω =   average angular velocity in rad/s
    θ =   angular displacement in radians
    t =   time taken in seconds

$$\omega = \frac{\theta}{t} \frac{\text{rad}}{\text{s}}$$

As noted in the previous section, because velocity is a vector quantity, the direction of rotation (either clockwise or anti-clockwise) must be quoted. If the direction of rotation is omitted, then the quantity being expressed is angular speed.

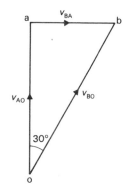

Figure 279   *Solution to self-assessment question 4*

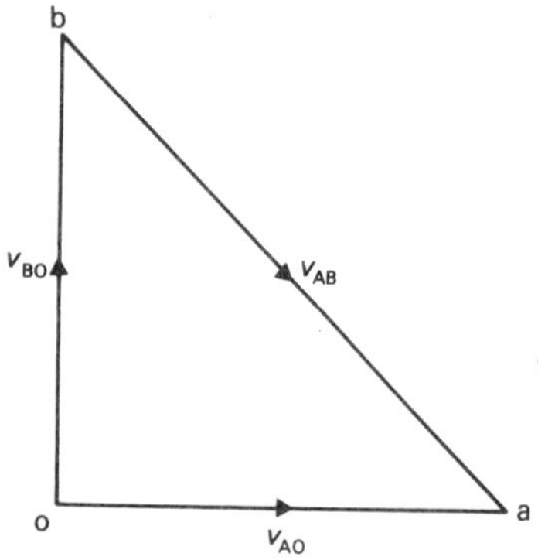

Figure 280   *Solution to self-assessment question 5*

Solutions to self-assessment questions

1   A 'speed limit', such as 50 km/hour, does not specify the direction in which a motor vehicle is travelling – indeed the direction of the vehicle could change frequently while it is travelling at 50 km/hour. Under these conditions, the vehicle's velocity changes every time its direction changes, whereas its speed remains constant. Thus 50 km/hour is correctly described as a speed limit.

2   The only vector quantity is velocity. Mass, efficiency and speed possess magnitude only, and are therefore scalar quantities. It might appear that pressure is a vector quantity, but since it acts in every direction (not a specified direction), it is a scalar quantity.

3   The correct alternative is (ii); the velocity of car B relative to car A is 220 km/hour due north.

4   The vector diagram is shown in Figure 279. $\overleftarrow{\text{ao}}$ represents the velocity of ship A relative to the beacon; $\overleftarrow{\text{bo}}$ represents the velocity of ship B relative to the beacon. The velocity of ship B relative to ship A is v_{BA}, and is represented by the vector $\overrightarrow{\text{ab}}$, i.e.

v_{BA} = 9.0 km/hour at an angle of 93° to the direction of motion of ship A.

5   The vector diagram is shown in Figure 280. $\overrightarrow{\text{oa}}$ represents the velocity of aircraft A relative to a fixed point, and $\overrightarrow{\text{ob}}$ represents the velocity of aircraft B relative to a fixed point. The velocity of aircraft A relative to B is v_{AB} and is represented by the vector $\overrightarrow{\text{ba}}$, i.e.

v_{AB} = 1 060 km/hour at an angle of 42° south of east.

The velocity of aircraft B relative to aircraft A is opposite in direction to that of A relative to B.

6   Assume the datum for velocity to be a fixed point at (say) the top of the escalator. The nett velocity of the athlete, relative to the fixed point is (8.5 − 0.5) m/s, i.e. 8.0 m/s upward.

$$\text{average velocity} = \frac{\text{displacement}}{\text{time taken}}$$

$$\therefore \quad \text{time taken} = \frac{\text{displacement}}{\text{average velocity}}$$

Angular acceleration

The strict definition of angular acceleration is *the rate of change of angular velocity with respect to time*. (As such it is exactly analogous to the definition of linear acceleration.) However, for the present, it is sufficient to define angular acceleration more simply as the change in angular velocity in unit time, i.e.

$$\text{angular acceleration} = \frac{\text{change in angular velocity}}{\text{time taken for the change}}$$

Using symbols, if the change in angular velocity is $(\omega_2 - \omega_1)$ and occurs in time t, the angular acceleration α, is given by:

$$\alpha = \frac{\omega_2 - \omega_1}{t} \frac{\text{rad/s}}{\text{s}}$$

$$\therefore \quad \text{time taken} = \frac{24}{8.0} \frac{\text{m}}{\text{m/s}}$$
$$= 3 \text{ seconds}$$

7 The vector diagram is shown in Figure 281. $\vec{oc}$ represents the velocity of the crane across the workshop, relative to the floor; $\vec{cd}$ represents the velocity of the casting relative to the suspension point (i.e. the velocity of descent). The velocity of the casting relative to the floor is v_{DO}, and is represented by the vector $\vec{od}$, i.e. $v_{DO} = 0.32$ m/s at an angle of 51° to the vertical.

8 The vector diagram is shown in Figure 282. $\vec{oa}$ represents the velocity of the first helicopter relative to a fixed point; $\vec{ab}$ represents the velocity of the second helicopter relative to the first. The true velocity of the second helicopter is represented by the vector $\vec{ob}$, i.e.
$v_{OB} = 86$ km/hour at 10° north of east.

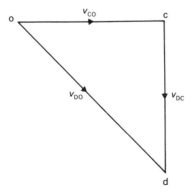

Figure 281 *Solution to self-assessment question 7*

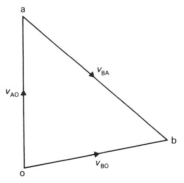

Figure 282 *Solution to self-assessment question 8*

Thus the unit of angular acceleration is rad/s^2, which is analogous to the unit of linear acceleration (m/s^2).

Almost all moving components in a machine experience accelerations at some stage in their operational cycle. Linear accelerations are used to determine some of the forces acting on components such as pistons, valves and crank pins. Similarly, angular acceleration is important since it is used to determine the forces acting on components such as the bearings of a shaft; it is also used in the calculation of the power transmitted by a clutch, or dissipated by a brake.

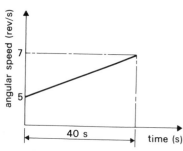

Figure 283   *Motion of a flywheel*

Example 3

A flywheel increases its speed uniformly from 5 to 7 rev/s in 40 seconds. Calculate the angular acceleration of the wheel.

In diagrammatic form, the motion of the flywheel is shown in Figure 283. The acceleration can be found from the equation:

$$\alpha = \frac{\omega_2 - \omega_1}{t} \text{ rad/s}^2$$

where $\omega_2 = 7$ rev/s $= 7 \times 2\pi$ rad/s $= 14\pi$ rad/s
(The arithmetic will be simplified if 14π is not yet evaluated as a number.)

$$\omega_1 = 5 \text{ rev/s} = 5 \times 2\pi \text{ rad/s} = 10\pi \text{ rad/s}$$

$$\alpha = \frac{\omega_2 - \omega_1}{t}$$

$$= \frac{14\pi - 10\pi}{40} \frac{\text{rad/s}}{\text{s}} = \frac{4\pi}{40} \text{ rad/s}^2 = 0.314 \text{ rad/s}^2$$

Equations of linear and angular motion
If the angular acceleration is constant, then using the same reasoning that is used for linear acceleration, the angular equivalent of the linear equations of uniform motion can be established.

Linear motion equations	*Angular motion equations*
$s = (\dfrac{v_1 + v_2}{2})t$	$\theta = (\dfrac{\omega_1 + \omega_2}{2})t$
$v_2 = v_1 + at$	$\omega_2 = \omega_1 + \alpha t$
$s = v_1 t + \tfrac{1}{2}at^2$	$\theta = \omega_1 t + \tfrac{1}{2}\alpha t^2$
$v_2^2 = v_1^2 + 2as$	$\omega_2^2 = \omega_1^2 + 2\alpha\theta$

where:

s	=	linear displacement,
v_1	=	initial linear velocity,
v_2	=	linear velocity at a later instant in time,
t	=	time for velocity to change change uniformly from v_1 to v_2,
a	=	linear acceleration.

θ	=	angular displacement,
ω_1	=	initial angular velocity,
ω_2	=	angular velocity at a later instant in time,
t	=	time for velocity to change uniformly from ω_1 to ω_2,
α	=	angular acceleration.

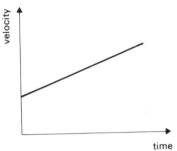

Figure 284   *Velocity time-graph*

Some readers may not be familiar with the third and fourth sets of equations. They can be derived, as can all the equations, from a velocity-time graph similar to that shown in Figure 284. Indeed, such a graph can be of considerable assistance in the solution of problems in either linear or angular motion. If the reader wishes, he is free to solve problems using only the equations given above, but the authors advocate the use of pictorial diagrams of velocity-time graphs as a useful aid in the analysis of both linear and angular motion.

The reader may remember that the slope of a uniform linear velocity-time graph yields linear acceleration. Using a similar argument, the slope of a uniform angular velocity-time graph is angular acceleration. Such an interpretation may occasionally save long, involved calculations.

Example 4

An electric motor, initially at rest, is caused to rotate with a uniform angular acceleration of 31.42 (i.e 10π) rad/s^2. Find, after 3s from the beginning of rotation, (i) the speed of rotation of the motor in rev/s, and (ii) the number of revolutions through which the motor has turned.

A sketch of the motion is shown in Figure 285. Using the equation:

$$\omega_2 = \omega_1 + \alpha t$$

where ω_2 is the angular velocity after 3 s

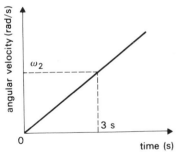

Figure 285   *Angular velocity-time*

$$\omega_1 = 0$$
$$\alpha = 10\pi \text{ rad/s}^2$$
$$t = 3 \text{ s}$$
$$\omega_2 = 0 + (10\pi \times 3)$$
$$= 30\pi \text{ rad/s} = \frac{30\pi}{2\pi} \text{ rev/s} = 15 \text{ rev/s}$$

The angular displacement can be found from the equation:

$$\theta = \left(\frac{\omega_1 + \omega_2}{2}\right)t$$

$$= \left(\frac{0 + 30\pi}{2}\right) 3 \text{ radians}$$

$$\therefore \; \theta \;\; = \; 45\pi \text{ rad} = \frac{45\pi}{2\pi} \text{ rev} = 22.5 \text{ revolutions}$$

Alternatively, the number of revolutions can be found using the equation:

$$\theta \;\; = \; \omega_1 t + \tfrac{1}{2}\alpha t^2$$

where $\omega_1 = 0$, $\alpha = 10\pi$ rad/s^2 and $t = 3$ s

$$\therefore \; \theta \;\; = \; 0 + (\tfrac{1}{2} \times 10\pi \times 9) = 45\pi \text{ rad}$$

Hence $\theta \;\; = \; \dfrac{45\pi}{2\pi} \text{ rev} = 22.5 \text{ revolutions.}$

The reader is free to choose which method he prefers.

Example 5

A flywheel is accelerated so that its velocity varies uniformly from rest; after 1 minute its speed of rotation is 5 rev/s. The flywheel runs at this speed for a further 5 minutes; it is then brought to rest in a uniform manner in a time of 50 seconds. Determine (i) the angular acceleration of the flywheel, (ii) the angular retardation of the flywheel, and (iii) the total number of revolutions made by the flywheel.

There are three distinct periods of motion of the flywheel, as illustrated in Figure 286.

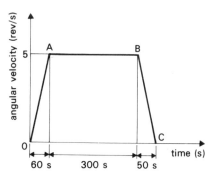

Figure 286   *Angular velocity-time for a flywheel*

From O to A, the flywheel is accelerating; the angular acceleration may be found graphically by plotting a graph of angular velocity (in rad/s) against time (in seconds). The slope of the resulting straight line is angular acceleration.

From A to B, the flywheel is rotating at constant angular velocity. Hence the slope of the line AB is zero, i.e. the acceleration is zero.

From B to C, the flywheel is decelerating; the deceleration may be found by plotting (in consistent units) angular velocity against time. (Note that the slope gives a negative quantity, indicating a deceleration.)

As an alternative to the graphical method, the acceleration and deceleration may be found in the following manner.

(i)   Considering the line OA in Figure 286

$$\omega_2 = 5 \text{ rev/s} = 5 \times 2\pi \text{ rad/s} = 10\pi \text{ rad/s},$$
$$\omega_1 = 0$$
$$t \;\; = 60 \text{ s}$$

Using $\omega_2 = \omega_1 + \alpha t$

$$10\pi = 0 + (\alpha \times 60)$$

$$\therefore \quad \alpha \quad = \quad \frac{10\pi}{60} \text{ rad/s}^2 = 0.524 \text{ rad/s}^2.$$

(ii) Considering the line BC in Figure 286,

$$\omega_1 = 10\pi \text{ rad/s}$$
$$\omega_2 = 0$$
$$t = 50 \text{ s}$$

Again, using $\omega_2 = \omega_1 + \alpha t$

$$0 = 10\pi + 50\alpha$$

$$\alpha = -\frac{10\pi}{50} \text{ rad/s}^2 = -0.628 \text{ rad/s}^2$$

The negative sign indicates a deceleration.

(iii) In order to determine the number of revolutions of the flywheel, each period of the motion is considered separately.
Using the equation:

$$\theta = (\frac{\omega_1 + \omega_2}{2}) t$$

$$\theta_{OA} = (\frac{0 + 10\pi}{2}) 60 \text{ radians}$$

$$= 300\pi \text{ rad}$$

$$\theta_{BC} = (\frac{10\pi + 0}{2}) 50 = 250\pi \text{ rad}$$

$$\theta_{AB} = \omega t \text{ (since } \omega_1 = \omega_2 = \omega)$$
$$\therefore \quad \theta_{AB} = 10\pi \times 300 = 3\,000\pi \text{ rad}$$

Total angle turned through is:
$$\theta = \theta_{OA} + \theta_{AB} + \theta_{BC}$$
$$= \pi(300 + 3\,000 + 250) \text{ rad}$$
$$= 3\,550\pi \text{ rad}$$

$$\text{Total number of revolutions} = \frac{3\,550\pi}{2\pi}$$

$$= 1\,775$$

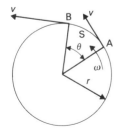

Figure 287 *Relationship between angular and linear velocity*

Relationship between angular and linear motion
Consider a crank pin moving from A to B in a circular path as shown in Figure 287. The pin is moving with constant angular velocity ω. Let the radius of the circular path be r metres, the angle between A and B be θ radians and the corresponding displacement round the circumference be s metres.

The instantaneous linear velocity of the crank pin is v m/s. Note that the direction of the linear velocity changes (at any given point it always acts tangentially to the circle), but its magnitude is constant. 1 radian is equivalent to a distance of r metres round the circumference.

Hence, θ radians is equivalent to a distance of θr metres round the circumference. But, the distance round the circumference subtended by the angle θ has been defined as s metres.

$$\therefore \quad s = \theta r \text{ or } \theta = \frac{s}{r}$$

If the crankpin moves from A to B in a time of t seconds, then:

$$v = \frac{s}{t} \text{ or } s = vt$$

$$\therefore \quad vt = \theta r$$

$$\frac{\theta}{t} = \frac{v}{r}$$

But, θ/t = angular velocity in radians/s,

$$\therefore \quad \omega = \frac{v}{r} \text{ or } v = \omega r$$

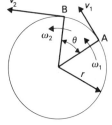

Figure 288   *Relationship between angular and linear acceleration*

Hence, for a disc moving with constant angular velocity ω, the linear velocity of a point on the circumference of the disc is directly proportional to the radius of the disc.

The analysis so far, has been restricted to constant angular velocity. Suppose the crankpin is moving with an angular acceleration. Referring to Figure 288, let the angular velocity increase uniformly in a time of t seconds from ω_1 at point A to ω_2 at point B; the corresponding increase in linear velocity of the crankpin is from v_1 to v_2. Let the linear acceleration around the circumference be a, so that:

$$v_2 = v_1 + at$$

$$\text{or } v_2 - v_1 = at$$

$$\text{But } v = \omega r$$

$$\therefore \quad \omega_2 r - \omega_1 r = at$$

$$\text{or } \quad \omega_2 - \omega_1 = \frac{at}{r}$$

$$\text{angular acceleration} = \frac{\text{change in angular velocity}}{\text{time taken for the change}}$$

$$\alpha = \frac{\omega_2 - \omega_1}{t}$$

$$\therefore \quad \alpha = \frac{at/r}{t}$$

$$\text{or} \quad \alpha \;=\; \frac{at}{r} \times \frac{1}{t} \;=\; \frac{a}{r}$$

$$\therefore \quad \alpha \;=\; \frac{a}{r} \quad \text{or} \quad a = \alpha r$$

Thus for a disc rotating at an angular acceleration α, the linear acceleration of a point on the circumference of the disc is directly proportional to the radius of the disc.

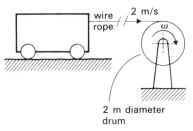

Figure 289   *Linear velocity converted to angular velocity*

Example 6

A truck is to be hauled along a horizontal surface at a constant velocity of 2 m/s by a wire rope wound onto a drum which is 2 m mean diameter. Calculate the angular speed at which the drum should be rotated.

Since the linear velocity of the truck is 2 m/s, the linear velocity of the wire must be 2 m/s. Using $\omega = v/r$ and referring to Figure 289,

$$\omega = \frac{2 \text{ m/s}}{1 \text{ m}} = 2 \text{ rad/s} = \frac{2}{2\pi} \text{ rev/s} = 0.318 \text{ rev/s} = 19.1 \text{ rev/min}$$

Example 7

A car travels at a constant speed of 72 km/hour for a distance of 1 km. If the wheels are 1 m diameter, calculate (i) the angular speed of the wheels in rev/s, and (ii) the number of revolutions turned through by each wheel.

If the car now accelerates uniformly to a speed of 108 km/hour in a time of 10 seconds, find the angular acceleration of the wheels in rad/s².

(i)   linear speed $= 72$ km/hour $= \dfrac{72 \times 10^3}{3\,600}$ m/s

$$\therefore v = 20 \text{ m/s}$$

angular speed of wheels is $\omega = \dfrac{v}{r} = \dfrac{20}{0.5} \dfrac{\text{m/s}}{\text{m}} = 40$ rad/s

angular speed of wheels in rev/s is $\dfrac{40}{2\pi} = 6.37$ rev/s

(ii)   time to travel 1 km $= \dfrac{\text{distance travelled}}{\text{constant linear velocity}}$

$$\therefore \quad \text{time to travel 1 km} = \frac{1\,000}{20}\frac{\text{m}}{\text{m/s}} = 50 \text{ s}$$

Hence, revolutions turned through in 1 km are:
6.37×50 rev $= 318.5$ revolutions

When the magnitude of the linear velocity of the car is

$$108 \text{ km/hour, i.e. } \frac{108 \times 10^3}{3\,600} = 30 \text{ m/s}$$

$$\text{angular velocity of wheels} = \frac{30}{0.5} \frac{\text{m/s}}{\text{m}} = 60 \text{ rad/s}$$

$$\text{angular acceleration of wheels} = \frac{\text{change in angular velocity}}{\text{time taken for change}}$$

$$\therefore \quad \text{angular acceleration} = \frac{60 - 40}{10} = 2 \text{ rad/s}^2$$

Alternatively, using $\alpha = \dfrac{a}{r}$

$$\text{linear acceleration} = \frac{30 - 20}{10} = 1 \text{ m/s}^2$$

$$\alpha = \frac{1}{0.5} \frac{\text{m/s}^2}{\text{m}} = 2 \text{ rad/s}^2$$

Self-assessment questions

9   Complete the following statement:
   'Angular velocity is defined as the rate of change of _____.'

10   1 radian is equivalent to:
   (i) 2π revolutions
   (ii) 1 revolution

   (iii) $\dfrac{1}{2\pi}$ revolutions

   (iv) $\dfrac{2\pi}{60}$ revolutions

   Select the correct alternative.

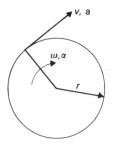

11   Referring to Figure 290, if linear velocity and acceleration are represented by v and a respectively, and angular velocity and acceleration by ω and α respectively, write down the relationship between v and ω, and between a and α.

12   The speed of a flywheel falls uniformly from 3 to 2 rev/s in 45 seconds. Determine the deceleration and the number of revolutions of the wheel over the period.

Figure 290   *Self-assessment question 11*

13   The drive shaft of a marine turbine, rotating at 6 rev/s, is uniformly retarded at a rate of 1.5 rad/s^2 for a time of 10 seconds. Calculate:

(i) the speed of rotation of the shaft in rev/s at the end of the retardation,

(ii) the number of revolutions of the shaft made in this time.

14 The shaft of an electric motor is accelerated uniformly in a clockwise direction from rest at a rate of 10 rad/s^2. After 30 seconds, the acceleration ceases, and the angular velocity is maintained constant for 1 minute; the velocity then falls uniformly to zero in a time of 25 seconds. Find:

(i) the maximum angular velocity of the shaft in rad/s,

(ii) the deceleration of the shaft in rad/s^2,

(iii) the total number of revolutions of the shaft.

15 The speed fluctuation of a flywheel is ± 2% about the mean speed of 4 rev/s. If the change from maximum to minimum speed (or vice-versa) takes place uniformly in a time of 40 seconds, calculate:

(i) the angular acceleration or deceleration of the wheel in rad/s^2,

(ii) the angular displacement (in radians) during the change.

16 A grinding wheel rotates at 40 rev/s. What is the cutting speed in metres per second if the wheel diameter is 180 mm?

17 A heavy mass is being pulled along a level floor at a speed of 0.2 m/s by supporting the mass on rollers 50 mm diameter. Find the speed of the rollers in rev/s, assuming that there is no slip between the mass and the rollers.

18 A boat travels at a constant speed of 36 km/hour round a curve of 500 m radius. Determine the time taken for it to turn through 180° of arc.

Solutions to self-assessment questions

9   The statement should be completed as follows:
'Angular velocity is defined as the rate of change of angular displacement with respect to time.'

This is a strict definition of angular velocity. For the purpose of this unit, the following relationship is sufficient:

$$\text{angular velocity} = \frac{\text{change in angular displacement}}{\text{time taken for the change}}$$

10   The correct alternative is (iii).

11   The relationships are:

$$v = \omega r \quad \text{or} \quad \omega = \frac{v}{r}$$

$$\text{and } a = \alpha r \quad \text{or} \quad \alpha = \frac{a}{r}$$

12   The behaviour of the flywheel is illustrated in Figure 291.
Using the relationship:

$$\alpha = \frac{\omega_2 - \omega_1}{t}$$

where $\omega_1 = 3 \times 2\pi$ rad/s $= 6\pi$ rad/s
$\omega_2 = 2 \times 2\pi$ rad/s $= 4\pi$ rad/s
and $t = 45$ s

$$\text{angular deceleration } \alpha = \frac{4\pi - 6\pi}{45} = -0.1396 \text{ rad/s}^2$$

$$\text{angular displacement } \theta = (\frac{\omega_2 + \omega_1}{2}) t$$

$$= (\frac{6\pi + 4\pi}{2}) 45 \text{ radian}$$

$$= \frac{5\pi \times 45}{2\pi} \text{ rev}$$

$\therefore$ angular displacement $= 122.5$ revolutions.

Alternatively, angular displacement can be found from:
$\theta = \omega_1 t + \frac{1}{2}\alpha t^2$
remembering that α is a negative quantity.

13   A sketch of the motion is illustrated in Figure 292.
(i)   Using $\omega_2 = \omega_1 + \alpha t$
where $\alpha = -1.5$ rad/s^2
$\omega_1 = 6 \times 2\pi$ rad/s $= 37.7$ rad/s
and $t = 10$ s
$\omega_2 = 37.7 - (1.5 \times 10) = 22.5$ rad/s

$$\therefore \text{ speed of rotation } = \frac{22.5}{2\pi} = 3.58 \text{ rev/s}$$

(ii)   angular displacement $\theta = (\frac{\omega_2 + \omega_1}{2}) t$

$$= (\frac{22.5 + 37.7}{2}) 10 \text{ rad} = 301 \text{ rad}$$

$$\therefore \text{ number of revolutions } = \frac{301}{2\pi} = 47.9$$

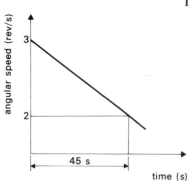

Figure 291   *Solution to self-assessment question 12*

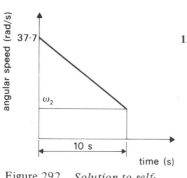

Figure 292   *Solution to self-assessment question 13*

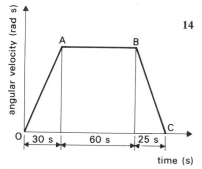

Figure 293   *Solution to self-assessment question 14 -*

14   The behaviour of the shaft is illustrated in Figure 293.
(i)   For the period O–A:

$$\omega_1 = 0$$
$$t = 30 \text{ s}$$
$$\alpha = 10 \text{ rad/s}^2$$

Using $\omega_2 = \omega_1 + \alpha t$
$$\omega_2 = 0 + (10 \times 30) = 300 \text{ rad/s}$$

∴ maximum velocity of shaft = 300 rad/s clockwise.

(ii)   For the period B–C:

$$\omega_1 = 300 \text{ rad/s}$$
$$\omega_2 = 0$$
$$t = 25 \text{ s}$$

Again,
using $\omega_2 = \omega_1 + at$
$$0 = 300 + 25\alpha$$

$$\alpha = -\frac{300}{25} \text{ rad/s}^2 = -12 \text{ rad/s}^2$$

∴ deceleration of shaft = 12 rad/s² clockwise.

(iii)   For the period O–A:

angular displacement $\theta_{OA} = (\dfrac{\omega_1 + \omega_2}{2}) t$

∴ $\theta_{OA} = (\dfrac{0 + 300}{2})$ 30 rad = 4 500 rad clockwise

For the period A–B:
angular displacement = ωt
∴ $\theta_{AB} = (300 \times 60)$ rad = 18 000 rad clockwise

For the period B–C:

angular displacement = $(\dfrac{\omega_2 + \omega_1}{2}) t$

∴ $\theta_{BC} = (\dfrac{300 + 0}{2})$ 25 rad = 3 750 rad clockwise

Total angular displacement
$$= \theta_{OA} + \theta_{AB} + \theta_{BC}$$
$$= (4\,500 + 18\,000 + 3\,750) \text{ rad}$$
$$= 26\,250 \text{ rad clockwise}$$

Total number of revolutions $= \dfrac{26\,250}{2\pi} = 8\,356$ clockwise

15   (i)   Mean speed of flywheel = $4 \times 2\pi$ rad/s = 8π rad/s

$$2\% \text{ of } 8\pi = \frac{2}{100} \times 8\pi = 0.16\pi \text{ rad/s}$$

∴   maximum speed of flywheel
$$= \omega_2 = (8 + 0.16)\pi \text{ rad/s} = 8.16\pi \text{ rad/s}$$
minimum speed of flywheel
$$= \omega_1 = (8 - 0.16)\pi \text{ rad/s} = 7.84\pi \text{ rad/s}$$

angular acceleration = $\alpha = (\dfrac{\omega_2 - \omega_1}{t})$

$$= \frac{(8.16\pi - 7.84\pi)}{40} \text{ rad/s}^2 = 0.025 \text{ rad/s}^2$$

Solutions to self-assessment questions

$$\text{angular displacement} = \left(\frac{\omega_1 + \omega_2}{2}\right)t$$

$$= \frac{(8.16\pi + 7.84\pi)}{2}\ 40\ \text{rad} = 16\pi \times 20\ \text{rad} = 1\ 005\ \text{rad}$$

16   Angular speed of grinding wheel $= 40 \times 2\pi$ rad/s
$\qquad\qquad\qquad\qquad\qquad\qquad\qquad\qquad\quad = 80\pi$ rad/s
The cutting speed in metres per second must be the linear velocity of a point on the rim of the wheel.
Hence, using $v = \omega r$
cutting speed $= (80\pi \times 0.09)$ m/s $= 22.62$ m/s

17   The arrangement of the rollers and heavy mass is shown in Figure 294(*a*). The linear velocity of the point of contact of the rollers and mass must be the same as the linear velocity of the mass.

Hence, referring to Figure 294 (*b*)

$$\omega = \frac{v}{r} = \frac{0.2}{0.025} = 8\ \text{rad/s}$$

$\therefore$ speed of rollers $= 8 \times 1/2\pi$ rev/s $= 1.273$ rev/s

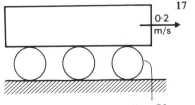

0·2 m/s

rollers 50 mm diameter

(a)

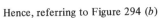

v = 0·2 m/s

ω

r

(b)

Figure 294   *Solution to self-assessment question 17*

18   The course of the ship is illustrated in Figure 295.

$$\text{Linear speed of ship} = 36\ \text{km/hour} = \frac{36 \times 10^3}{3\ 600}\ \text{m/s} = 10\ \text{m/s}$$

$$\text{angular velocity of ship} = \omega = \frac{v}{r} = \frac{10}{500} = 0.02\ \text{rad/s}$$

angular displacement, $\theta = \omega t$
where t is the time for the displacement to occur.
But angular displacement $= 180° = \pi$ rad

$$t = \frac{\theta}{\omega} = \frac{\pi}{0.02}\ \frac{\text{rad}}{\text{rad/s}} = 157\ \text{s}$$

An alternative solution is as follows:
Circumference of 180° arc $= 500\pi$ metres

$$\text{linear speed of ship} = \frac{\text{linear distance travelled}}{\text{time taken}}$$

$$\therefore\ \text{time taken} = \frac{\text{distance travelled}}{\text{speed}}$$

$$= \frac{500\pi}{10}\ \frac{\text{m}}{\text{m/s}} = 157\ \text{s}$$

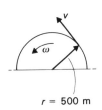

v

ω

r = 500 m

Figure 295   *Solution to self-assessment question 18*

Section 3

Newton's laws of motion

After reading the following material, the reader shall:

3   Appreciate the effects of inertia and be able to relate force, mass and acceleration.

3.1   State Newton's first law of motion.

3.2   State Newton's second law of motion in terms of $F = ma$.

3.3   Explain mass in terms of inertia.

3.4   Solve problems using $F = ma$.

3.5   Define the newton.

3.6   State that when a force acts on one body there must be an equal and opposite force on another body.

Dynamics is concerned with motion. The fundamentals of the science of dynamics were discovered by Galileo in the seventeenth century. However, it was not until later in that century, in 1687, that Sir Isaac Newton summarised the then present knowledge in the form of three laws. These laws still remain fundamental to present day analysis of dynamical systems.

Newton's first law of motion: Every body (or object) remains in a state of rest, or continues in a state of uniform motion in a straight line, unless it is compelled to change that state by an externally applied force.

This principle is used in spaceflight, remote from the gravitational field of the earth or any other planet; a space ship will continue to travel at constant velocity, even though its engine is switched off.

Newton's second law of motion: If an external force is applied to a body so that the body is caused to accelerate in the direction of the force, then the product of the mass and acceleration of the body is proportional to the applied force.

Using F as the symbol for force

           m as the symbol for mass

and a as the symbol for acceleration,

           $F \propto ma$

In contrast to the first law, Newton's second law is concerned with bodies whose velocity is changing, i.e. a body is subjected to a force which causes an acceleration.

Newton's third law of motion: When a force acts on one body, there must be an equal and opposite force on another body. A simplified version of the third law (*To every force there is an equal and opposite reaction*) has already been utilised in the topic area *statics*; it can also be used as an aid to the investigation of topics in dynamics, but detailed discussion of its use will be postponed until later in this topic area. For the moment, the discussion is restricted to Newton's first and second laws.

Consider a stationary motor vehicle on a level, straight road. Newton's first law asserts what may well be familiar to the reader, that the vehicle will remain stationary unless an external force is applied to it (the force may take any form: an extremely strong gust of wind, willing helpers to push and pull, or it may originate from the engine). Once the external force is applied, Newton's second law asserts that the vehicle will accelerate; the force is proportional to the product of the mass and acceleration of the vehicle. Also, the vehicle will accelerate in the direction of the applied force.

Instead of a stationary vehicle, consider now the case of a vehicle travelling at constant velocity along a level, straight road. Newton's first law maintains that the vehicle will continue to travel at constant velocity until an external force is applied to it. If the velocity changes, this means that the vehicle is accelerating. This acceleration must be caused by a force. According to Newton's second law, the magnitude of the external force causing this acceleration is proportional to the product of mass and acceleration.

Hence, if an external force is applied, either to a stationary body or to a body moving with constant velocity, the relationship between the force and the mass and ensuing acceleration of the body is:
$F \propto ma$

The proportional sign $(\propto)$ can be replaced by an equals sign $(=)$ providing a constant of proportionality is introduced, e.g.
$F = ma \times$ constant

In earlier sections of this book, the newton (symbol N) has been used as the S.I. unit of force. In S.I. one newton is defined as *the force required to give a mass of 1 kg an acceleration of 1 m/s^2*.

Substituting $F = 1$ N, $m = 1$ kg and $a = 1$ m/s^2 in the above expression gives:

$$1[N] = 1[kg] \times 1[\frac{m}{s^2}] \times \text{a constant}$$

Hence the constant is equal to unity. Thus, providing that the force F is in newtons, the mass m is in kg and the acceleration a is in metres per second per second, then:
$F = ma$

Note that the unit of force can be expressed in terms of the units of mass, length and time, e.g.

1 newton is equivalent to 1 kg × 1 m/s^2

or N is equivalent to $\dfrac{\text{kg m}}{\text{s}^2}$

Acceleration due to gravity

If an object is dropped from a high building, or an aeroplane, the velocity of the object increases as it falls, up to a certain value, depending upon air resistance. Thus, the object accelerates. It might appear that, not only the velocity is increasing as the object falls, but also that the acceleration is increasing. In fact, this is not so; an object in 'free-fall' is subjected to a constant acceleration throughout the period of free-fall. Because the earth is not a perfect sphere, the free-fall acceleration varies in magnitude at different points on the earth's surface. However, at any given point on the surface of the earth, the acceleration in free-fall is constant. This acceleration is called *the acceleration due to gravity*; it is denoted by the symbol *g*, and may be assumed to have a magnitude of 9.81 m/s^2. (In London at sea level, *g* is almost exactly 9.81 m/s^2; at the equator, *g* is approximately 9.78 m/s^2; at the poles, *g* is approximately 9.832 m/s^2).

Thus, an object falling to earth experiences an acceleration which can be assumed to be 9.81 m/s^2. The object has mass. Hence the object has a force applied to it. This force is different from most other forces in that its direction never changes — it *always* acts towards the centre of the earth. Because it acts in one direction only, it is given a distinguishing name — *weight*. In other words, weight is the gravitational force exerted by every object which has mass.

Consider a mass of 1 kg resting on a table; the force exerted by the mass on the table is the gravitational force (i.e. its weight), since if the table is removed, the mass accelerates at approximately 9.81 m/s^2. The magnitude of the mass is known, as is the magnitude of the acceleration, and therefore the magnitude of the gravitational force can be calculated. Remembering that:

$$F\,[\text{N}] \;=\; m\,[\text{kg}] \times a\,[\text{m/s}^2]$$

and substituting *m* = 1 kg and *a* = 9.81 m/s^2,

$$F\,[\text{N}] \;=\; 1 \times 9.81\,[\text{kg m/s}^2]$$
i.e. $F \;=\; 9.81\ \text{N}$

Hence, the weight of 1 kg is 9.81 N, and in more general terms:
weight = mass × *g*

Inertia force

If a body is in static equilibrium, then using Newton's third law, the

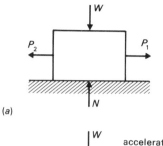

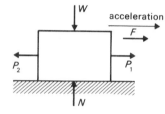

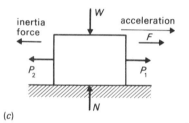

Figure 296   *Inertia force*
producing static equilibrium

applied forces are 'balanced' by equal and opposite forces. Thus in Figure 296 (*a*), the applied forces of W and P_1 are balanced by equal and opposite forces N and P_2.

i.e. $W = N$

and $P_1 = P_2$

Applying Newton's first law, the same system of forces exists when the body is moving at constant velocity.

If a body is accelerating, then using Newton's first and second laws, there must be a net force which is not balanced by an equal and opposite force. The body accelerates in the direction of the net force, which is of magnitude:

$F = ma$

The system of forces acting on the body may then be as shown in Figure 296 (*b*). The system is no longer in static equilibrium.

The concept of force used in dynamics is the same as that used in statics; hence it is useful to be able to apply the principle of static equilibrium to the analysis of dynamic problems. In order to restore static equilibrium to Figure 296 (*b*), an imaginary force of magnitude *ma* is introduced, acting in the opposite direction to the accelerating force. This imaginary force is called the *inertia force*; its direction is *always* opposite to that of the acceleration. Figure 296 (*c*) shows the complete system of forces acting on the body, which is now in static equilibrium.

Remember that inertia force does not actually exist – if it did, a body could never accelerate. However, it is a useful device in the solution of problems in dynamics because it transforms a dynamic problem into an easier to solve static problem.

Inertia force is equal to the product of mass and acceleration, and is therefore directly proportional to mass. Thus, it can sometimes be helpful to think of mass in terms of inertia, i.e. to think of mass in terms of a reluctance to move or to be accelerated; this idea should readily be appreciated by the reader if he reflects upon the relative difficulty of moving a large mass in comparison with a smaller mass.

Example 8
A mass of 1 kg is hung from a spring balance in a lift. Determine the spring balance reading when the lift is (i) at rest, (ii) moving upwards with an acceleration of 3 m/s², (iii) moving downwards with an acceleration of 3 m/s², and (iv) moving downwards with a deceleration of 3 m/s². Assume the acceleration due to gravity is 9.81 m/s².

(i)   When the lift is at rest, the spring balance registers the gravitational force acting on the mass.

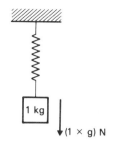

(a)

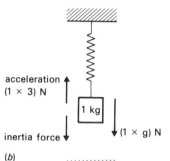

(b)

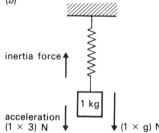

(c)

Figure 297 *Directions of accelerating and inertia forces*

Using $F = ma$, and referring to Figure 297 (*a*),
$$F = (1 \times 9.81)\,N = 9.81\,N \text{ downwards}$$
The spring balance reading is 9.81 N.

(ii) When the lift is accelerating upwards at 3 m/s^2, the forces acting are shown in Figure 297 (*b*).
inertia force = $ma = 1 \times 3 = 3\,N$
This opposes the accelerating force, i.e. it acts downwards.
Hence, total downward force = $(9.81 + 3)\,N$
∴ spring balance reading = 12.81 N

(iii) When the lift accelerates downward at 3 m/s^2, the forces acting are shown in Figure 297 (*c*).
inertia force = $(1 \times 3)\,N = 3\,N$ acting upward
∴ spring balance reading = $(9.81 - 3)\,N = 6.81\,N$

(iv) A deceleration is a negative acceleration. Hence, when the lift is moving *downwards* with a *deceleration* of 3 m/s^2, it can be assumed to be *accelerating upwards* at 3 m/s^2. This condition is the same as case (ii).
∴ spring balance reading = 12.81 N

Example 9
The forge head of a horizontal forging machine has a mass of 750 kg. The velocity of the head at the instant it touches the block being forged (i.e. the billet) is 2 m/s, and the head is brought to rest in a distance of 25 mm by the uniform resistance of the block. Calculate the resisting force.

The deceleration of the head can be found from the equation:
$$v_2^2 = v_1^2 + 2as$$
where $v_2 = 0$
        $v_1 = 2\,\text{m/s}$
and $s = 0.025\,\text{m}$
∴ $0 = 4 + 0.05a$

$$a = -\frac{4}{0.05} = -80\,\text{m/s}^2$$

Using $F = ma$
retarding force = $[750 \times (-80)]\,N$
               = $-60\,000\,N = -60\,kN$
∴ resisting force = $60\,kN$

Note the very high value of resistance offered by the billet being forged. It has important implications for the designer of the forging and the designer of the forging machine.

Action and reaction
It is often necessary to distinguish between action and reaction, when these words are used to describe forces. An *active* force is one which

causes or tends to cause a change in velocity, e.g. the force due to gravity or the force from the engine of a car when the accelerator pedal is depressed. A *reactive* force occurs in response to the effects of an active force; examples include the recoil of a gun, the upward reaction of a beam support and inertia force.

Action and reaction are two equal and opposite forces exerted between two separate bodies. For example, if a mass of 20 kg rests on a planing machine table, it exerts an active force of (20×9.81) N on the table; the table exerts an equal and opposite reaction on the mass.

Example 10
A rocket in level flight ejects 200 kg of burnt fuel at a velocity of 500 m/s relative to the rocket. All of the fuel is ejected in a time of 10 seconds. Calculate the reaction on the rocket due to the ejection of the fuel, assuming that the relative velocity of the fuel increases uniformly.

The acceleration of the fuel relative to the rocket can be calculated from:

$$v_2 = v_1 + at$$

where v_2 and v_1 are velocities of the fuel relative to the rocket. The initial velocity of the fuel relative to the rocket is zero, i.e. $v_1 = 0$ (see Figure 298 (*a*)).

$$v_2 = 500 \text{ m/s}$$
$$\text{and} \quad t = 10 \text{ s}$$
$$\therefore \quad 500 = 0 + (a \times 10)$$
$$\therefore \quad a = 50 \text{ m/s}^2$$

The accelerating force on the fuel (i.e. the active force) is
$$F = ma = 200 \times 50 = 10 \text{ kN}$$

The reaction on the rocket is equal and opposite to this (see Figure 298 (*b*)).

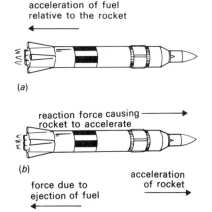

acceleration of fuel
relative to the rocket

(a)

reaction force causing
rocket to accelerate

(b)

force due to
ejection of fuel

acceleration
of rocket

Figure 298   *Action and reaction*

Self-assessment questions

19   State Newton's first law of motion.

20   State Newton's second law of motion in terms of the mass and acceleration of a body.

21   Complete the following statement:
'When a force acts on one body, there must be an equal and opposite _____.'

22   Select the correct alternative, only one of which is correct, from the list below. (The symbol ≡ means 'equivalent to'.)

(i) $N \equiv kg/m/s^2$
(ii) $N \equiv kg \ m/s^2$
(iii) $N \equiv kg/m/s$
(iv) $N \equiv kg \ m/s$

23   Explain how the mass of a body is related to its inertia, making reference if necessary to Newton's laws of motion.

24   When a mass of 1 kg is hung on a very accurate spring balance, the reading is 9.809 N. If the mass and spring balance are removed to another geographical location where the acceleration due to gravity is 9.832 m/s², what is the difference in the reading of the spring balance compared with the first reading? What is the acceleration due to gravity at the first geographical location? Quote both answers to 3 decimal places.

25   A train has a mass of 220 tonnes (1 tonne = 1 000 kg). Starting from rest, it accelerates uniformly to a velocity of 72 km/hour in 40 seconds. Calculate the force causing acceleration.

26   A loaded lorry has a mass of 5 000 kg and is travelling along a straight, level road at 36 km/hour. When the brakes are applied, its velocity falls uniformly to zero in a distance of 25 m. Determine the braking force.

27   A rocket ejects 3 000 kg of burnt gases at a velocity of 250 m/s relative to the rocket. Calculate the force on the rocket, if the velocity of the gases increases uniformly from zero to 250 m/s in a time of 5 seconds.

28   The head of a vertical forging machine has a mass of 5 000 kg. It falls freely from rest over a distance of 2 m, and after striking the billet, is brought to rest uniformly in a distance of 100 mm. Calculate the impact velocity and hence the resistance of the billet to the head. Assume acceleration due to gravity to be 9.81 m/s².

29   A lift of mass 300 kg has its maximum acceleration at the beginning of its ascent of 0.5 m/s². If the total cross-sectional area of the hoisting cables is 400 mm², calculate the maximum stress in the cables. Assume the acceleration due to gravity is 9.81 m/s².

$$\left(\text{Note that tensile stress} = \frac{\text{tensile force}}{\text{cross-sectional area}} \cdot \right)$$

Solutions to self-assessment questions

19   Newton's first law of motion may be written as:
'Every object remains in a state of rest, or continues in a state of constant velocity, unless it is compelled to change that state by an externally applied force.'

20   Newton's second law of motion may be written as:
'The force causing a body to accelerate is proportional to the product of the mass of the body and the acceleration of the body. The body accelerates in the direction of the applied force.'
Using symbols, this statement means:
$F \propto ma$

Note that it is not really correct to write $F = ma$, unless consistent units are chosen for the three variables.

21   The statement should be completed as follows:
'When a force acts on one body, there must be an equal and opposite force acting on another body.'
The completed statement is a practical interpretation of Newton's third law of motion.

22   The correct alternative is (ii), i.e. $N \equiv kg\ m/s^2$.

23   The mass of an accelerating body is directly proportional to the force causing acceleration. Using Newton's third law of motion, mass is therefore directly proportional to inertia force. Hence mass can be thought of as a reluctance to be accelerated; that is, mass can be considered as the inertia to be overcome before movement can begin.

24   One of the principles of operation of a spring balance is Newton's second law of motion.
Using $F = ma$
where F is the spring balance reading, and a the acceleration due to gravity:
$F = 1 \times 9.832 = 9.832\ N$
$\therefore$ difference in readings $= 0.023\ N$
When the spring balance reading is 9.809 N, then

$$a = \frac{F}{m} = \frac{9.809}{1} = 9.809\ m/s^2$$

25   The acceleration of the train can be calculated from the equation:
$$v_2 = v_1 + at$$

where $v_2 = 72\ km/hour = \dfrac{72 \times 10^3}{3\,600}\ m/s = 20\ m/s$

$v_1 = 0$
$t = 40\ s$
$\therefore 20 = 0 + (a \times 40)$

$$a = \frac{20}{40} = 0.5\ m/s^2$$

Accelerating force $= ma$
$\qquad\qquad\qquad = 220 \times 10^3 \times 0.5\ N = 110\ kN$

26   The deceleration of the lorry can be found from the equation:
$$v_2{}^2 = v_1{}^2 + 2as$$
where $v_2 = 0$

$$v_1 = 36 \text{ km/hour} = \frac{36 \times 10^3}{3\,600} \text{ m/s} = 10 \text{ m/s}$$

$$s = 25 \text{ m}$$
$$\therefore \quad 0 = 100 + (2 \times 25 \times a)$$
$$a = -\frac{100}{50} = -2 \text{ m/s}^2$$

Braking force $= ma$
$$= 5\,000 \times 2 = 10 \text{ kN}$$

27   The acceleration of the gases relative to the rocket can be determined from the equation:

$$v_2 = v_1 + at$$

where v_2 and v_1 are velocities of the gases relative to the rocket. The initial velocity of the gases relative to the rocket is zero, i.e.

$$v_1 = 0$$
$$v_2 = 250 \text{ m/s and } t = 5 \text{ s}$$
$$\therefore \quad 250 = 0 + (5a)$$
$$a = \frac{250}{5} = 50 \text{ m/s}^2$$

The accelerating force on the gases is found from
$$F = ma = 3\,000 \times 50 = 150 \text{ kN}$$

The force on the rocket (i.e. the reaction force) is equal and opposite to this.

28   The impact velocity of the head of the machine is found from the equation:
$$v_2{}^2 = v_1{}^2 + 2as$$
where $v_2 = $ impact velocity
$$v_1 = 0$$
$$a = 9.81 \text{ m/s}^2$$
$$s = 2 \text{ m}$$
$$\therefore \quad v_2{}^2 = 0 + (2 \times 9.81 \times 2) = 39.24 \text{ (m/s)}^2$$
$$\therefore \quad v_2 = 6.26 \text{ m/s}$$

The deceleration of the head of the machine as it deforms the billet is found from the equation:
$$v_2{}^2 = v_1{}^2 + 2as$$
$$v_2 = 0 \text{ (i.e. at the end of the deformation, the forging head is stationary)}$$
v_1 is now 6.26 m/s
$$s = 100 \text{ mm} = 0.1 \text{ m}$$
$$\therefore \quad 0 = 39.24 + 0.2a$$
$$a = -\frac{39.24}{0.2} = -196.2 \text{ m/s}^2$$

Retarding force $= ma = (5\,000 \times 196.2) \text{ N} = 981 \text{ kN}$
In addition to the retarding force, there is another force helping to deform the forging. This is the force due to gravity of the forging machine head.
weight of head $= mg = (5\,000 \times 9.81) \text{ N} = 49.05 \text{ kN}$
Hence, total resistance of the billet is:
$(981 + 49.05) \text{ kN} \simeq 1\,030 \text{ kN or } 1.03 \text{ MN}$

Solution to self-assessment question

29   Maximum accelerating force on the lift is:

$F = ma = (300 \times 0.5)\,\text{N} = 150\,\text{N}$

This force acts upwards, and hence the maximum inertia force of 150 N acts downwards. The additional force to be borne by the cables is the weight of the lift. Weight of lift = $(300 \times 9.81)\,\text{N} = 2\,943\,\text{N}$ acting downwards. Hence, the total force to be borne by the cables is:

$(2943 + 150)\,\text{N} = 3093\,\text{N}$

Maximum stress in the cables is:

$$\frac{3\,093}{400}\ \frac{\text{N}}{\text{mm}^2} = \frac{3\,093}{400 \times 10^{-6}}\ \frac{\text{N}}{\text{m}^2}$$

$$= \frac{3\,093 \times 10^6}{400}\ \frac{\text{N}}{\text{m}^2} = \frac{3\,093}{400}\ \frac{\text{MN}}{\text{m}^2} = 7.73\,\text{MN/m}^2$$

Section 4

Friction between horizontal surfaces

After reading the following material, the reader shall:

4   Appreciate the effect of friction between horizontal surfaces in contact.

4.1   State the laws of friction.

4.2   Distinguish between 'static' and 'dynamic' friction.

4.3   Solve problems involving dynamic friction.

4.4   Describe the effect of lubricating the two surfaces in contact.

If two surfaces are moving, or are on the point of moving, relative to each other, a force occurs which opposes the movement between the surfaces. The force is a result of friction between the surfaces, and is called *the friction force*.

There are occasions when engineers make use of friction forces; these include clutches, drum or disc brakes, belt drives or the clamping of a component prior to machining. Instances where it is desirable that the friction force should be as low as possible include bearings, gears or a slide and slider such as occurs with a shaping machine ram. All of these applications can be classified by at least one of the various categories of friction as outlined below:

category of friction	operating conditions and applications
dry friction	The surfaces are free from dirt, water or grease, e.g. clutches, brakes, belt drives.
fluid or viscous friction	The surfaces are separated by a film of fluid, so that friction depends wholly upon the lubricant. All bearings should operate under fluid-film lubrication.
pure rolling friction	The friction force under pure rolling conditions is very much less than under sliding conditions. Hence, in designing machines, sliding friction is often replaced by rolling friction, e.g. ball or roller bearings are used instead of plain bearings.

The following discussion is limited in the main to friction occurring under dry, clean conditions; the analysis of fluid and rolling friction is beyond the scope of this book. However, the effect of lubricating two surfaces in contact is briefly discussed later in this section.

Static friction

Consider an object of mass m resting on a horizontal table as shown in Figure 299. The object exerts a downward force on the table, the force

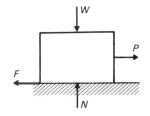

Figure 299   *Static friction*

being the weight of the object. Using Newton's third law of motion, the table exerts a force equal to and opposite to the weight. The weight is denoted by W in Figure 299, and the force equal and opposite to it — the *normal* force — is denoted by N ('normal' in this context means at 90° to the surface).

If a force P is applied, tending to slide the object across the table, it will be resisted by the friction force F. Note the phrase 'tending to slide the object across the table'; the object may be on the point of moving, but it is stationary. The friction which occurs under such conditions is termed *static* friction. The object is in static equilibrium and hence $W = N$ and $P = F$.

It is possible to conduct a series of experiments on apparatus similar to that depicted in Figure 299, using surfaces in contact which are dry and clean. The conclusions drawn from these experiments are often called the 'laws of friction'. To be strictly correct, they are not laws and are better classified as experimental observations; they may be summarized as follows:

1   The friction force always opposes motion or intended motion.
2   The friction force is very dependent upon the character and surface finish of the materials in contact.
3   The friction force is independent of the area of contact between the surfaces.
4   The friction force is directly proportional to the normal force between the surfaces.

It is re-emphasized that the observations apply only to dry, clean surfaces in contact.

Statement 4 can be expressed in 'shorthand' form, using the notation of Figure 299:
$$F \propto N$$
i.e. $F = N \times$ a constant

The constant of proportionality is usually denoted by the symbol μ (Greek letter mu), so that:
$$F = \mu N$$
μ is termed the *coefficient of friction.*

Since F has been defined as a friction force, and N is the normal force between the surfaces in contact, the coefficient of friction has no units, and is simply a ratio.

The value of the coefficient of friction depends upon the materials in contact; it also depends upon the nature of the two surfaces in contact. For example, if two steel surfaces are in contact, an average value of coefficient of friction is about 0.2; this value increases quite markedly if the surfaces in contact are rough; conversely, the value reduces if the surface finish is smooth. Typical average values of coefficient of friction under static conditions are shown in the table below.

materials	coefficient of friction
metal on metal	0.2
leather or rubber on metal	0.4
asbestos on cast iron	0.45
rubber on road surface	0.9

Note: the cleanliness and roughness of the surfaces involved have a marked effect upon the coefficient of friction. Average values are quoted, but widely different values can be obtained by (*a*) changing the surface finish of the surfaces, (*b*) using a lubricant between the surfaces in contact. This last point will be discussed in a little more detail later in this section.

Kinetic or dynamic friction

The discussion so far has been limited to the instant before an object begins to slide across a surface. At this instant, the value of the friction force is termed the *limiting friction force*, and the coefficient of friction is often called the coefficient of *static* friction.

The conclusions from the experiments on static friction are applicable to moving objects, *providing that the speed of motion is constant and relatively low*. In other words, there must be no accelerating force acting on the object. Thus, the equation connecting F, μ and N is applicable, providing that the surfaces are dry and clean, and that the speed is uniform and low.

There is one other proviso to be remembered when comparing the effects of kinetic friction with those of static friction: the coefficient of kinetic friction for two materials in contact is usually slightly less than the coefficient of static friction for the same two materials. Otherwise, kinetic or sliding friction obeys the same 'laws' as static or limiting friction.

Example 11

It is required to pull a mass of 1 000 kg along a horizontal surface by means of a steel rope. The coefficient of kinetic friction between the mass and the surface is 0.3. If the mass moves at a low uniform speed across the surface, calculate the horizontal pulling force required. Assume g as 9.81 m/s^2.

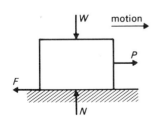

·Figure 300   *Kinetic friction*

The system of forces applied to the mass is shown in Figure 300.
Gravitational force acting on mass $= W = (1\,000 \times 9.81)$ N
$$= 9\,810 \text{ N}$$
$\therefore$ normal force $= N = 9\,810$ N

$$\text{friction force} = F = \mu N$$
$$= (0.3 \times 9\,810)\,\text{N}$$
$$= 2\,943\,\text{N}$$

Since the mass is moving at a low uniform speed (i.e. it is not accelerating), $P = F$.

$\therefore$ pulling force = $2\,943\,\text{N}$ = $2.943\,\text{kN}$

Lubrication of friction surfaces

The force required to move an object at uniform speed is dependent upon the friction force. Hence the required pulling force can be reduced if the friction force is reduced. A common method of reducing friction force is by ensuring that the friction surfaces are lubricated.

An everyday, and yet at times dramatic, example of the effects of lubrication can be observed by comparing the performance of a motor vehicle on a dry road, with its performance on a wet road. Compared with a dry surface, the distance required to stop a motor vehicle on a wet road increases dramatically; in other words, the coefficient of friction between the road surface and the tyres is reduced.

Suppose the amount of water on the road is such that the tyres make no direct contact with the road surface. The effect on the braking performance of the vehicle may be very significant, and occasionally catastrophic! The water *separates* the tyres from the road surface. It is this principle which is often used by the engineer when he attempts to minimise friction forces between moving parts.

When two surfaces are separated by a film of fluid (such as oil), the friction force depends wholly upon the lubricant, and not upon the nature or material of the surfaces. The friction force is that force required to shear the lubricant film; it is in general very much less than the corresponding friction force between two unlubricated surfaces.

Fluid friction occurs therefore when surfaces are separated by a film of fluid or lubricant. It exists only when there is motion between surfaces, since under static conditions, some of the lubricant would be squeezed out by the normal force between the surfaces. One of the most common applications of fluid friction is in bearings. One of the major functions of plain bearings is to bear relatively high forces whilst keeping friction forces to a minimum. If the applied force on the bearing is very high, or the speed of sliding is relatively low, the fluid film approaches what is known as *boundary lubrication*. The surfaces are still separated by a film of lubricant, but it is now a very thin layer, perhaps 0.001 mm thick. Under excessive bearing forces, the boundary layer itself may break down. Contact takes place between high spots on the metal surfaces; the resulting high temperatures may cause local melting and seizure.

Self-assessment questions

30 Complete each of the following statements:
(i)   The friction force always opposes_____.
(ii) The friction force is directly proportional to the_____.

31 Answer true or false to each of the following statements:
(i)   The 'laws of friction' are applicable only to dry clean surfaces.

TRUE/FALSE

(ii) The 'laws of friction' are applicable to lubricated surfaces, when fluid film lubrication exists.

TRUE/FALSE

(iii) When two surfaces are separated by a layer of lubricant, the friction force depends upon the materials of the surface.

TRUE/FALSE

(iv) Static friction refers to bodies which are stationary, or to bodies which are on the point of moving relative to each other.

TRUE/FALSE

(v)   Kinetic friction refers only to bodies which are moving relative to each other.

TRUE/FALSE

(vi) Fluid friction may occur under static conditions.

TRUE/FALSE

(vii) Under static friction conditions, the friction force is independent of the area of contact between the surfaces.

TRUE/FALSE

32 A rolled steel joist is pulled at constant speed along a horizontal floor by a chain which is parallel to the floor. If the coefficient of friction between the joist and the floor is 0.6, and the mass of the steel joist is 300 kg, determine the force in the chain. Assume $g = 9.81$ m/s^2.

33 The table of a planing machine has a mass of 600 kg and moves horizontally on its bed. If the coefficient of friction between the table and the bed is 0.07, find the force required to overcome friction when a casting of mass 800 kg is bolted on the machine table. Assume $g = 9.81$ m/s^2.

34 A mass of 1 000 kg is pulled along a horizontal track at uniform speed by a steel rope which is parallel to the track. If the coefficient of friction between the mass and the track is 0.2, and the cross-sectional area of the rope is 300 mm^2, find the resultant stress in the rope. Assume $g = 9.81$ m/s^2.

35 A horizontal force of 350 N is required to haul a loaded trolley of total mass 200 kg at uniform speed along a level track. If the load on the trolley is increased so that its total mass is 300 kg, calculate the horizontal pulling force required to move the trolley at uniform speed. Assume $g = 9.81$ m/s^2.

Solutions to self-assessment questions

30   The completed statements are:
    (i)   Friction force always opposes motion.
    (ii) Friction force is directly proportional to the normal force between the surfaces.

31   (i)   TRUE.
    (ii) FALSE; the 'laws of friction' apply only to surfaces which are clean and dry.
    (iii) FALSE; if the two surfaces are separated by a layer of lubricant, the friction force depends upon the properties of the lubricant, not upon the materials of the surfaces.
    (iv) TRUE.
    (v) TRUE.
    (vi) FALSE; fluid friction only occurs when the surfaces are moving relative to each other.
    (vii) TRUE.

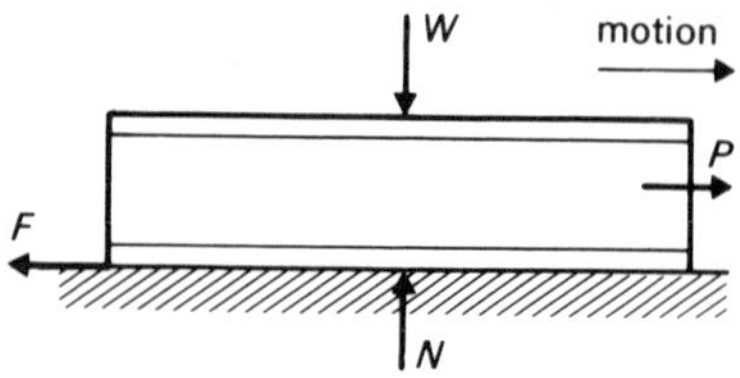

Figure 301   *Solution to self-assessment question 32*

32   The arrangement is shown diagrammatically in Figure 301.
$$W = mg = (300 \times 9.81)\ N = 2\,943\ N$$
$\therefore$ normal force $= N = 2\,943\ N$
friction force $= F = \mu N = (0.6 \times 2\,943)\ N = 1\,766\ N$
Since the joist is being pulled at uniform speed, $P = F$.
$\therefore$ force in the chain $= 1\,766\ N = 1.766\ kN$

33   Total mass $= (600 + 800)\ kg = 1\,400\ kg$
gravitational force $= (1\,400 \times 9.81)\ N = 13.73\ kN$
$\therefore$ normal force between surfaces $= N = 13.73\ kN$
friction force $= \mu N = 0.07 \times 13.73\ kN = 961\ N$
$\therefore$ force to overcome friction $= 961\ N$.

34   Gravitational force on mass $= W = (1\,000 \times 9.81)\ N = 9.81\ kN$
$\therefore$ normal force $= N = 9.81\ kN$
friction force $= F = \mu N = (0.2 \times 9.81)\ kN = 1.962\ kN$
Since the mass is being pulled at uniform speed, $P = F$
$\therefore P = 1.962\ kN$
Resultant stress in rope is given by
$$\frac{force}{cross\text{-}sectional\ area}$$
$\therefore$ resultant stress $= \dfrac{1.962}{300}\ \dfrac{kN}{mm^2} = \dfrac{1.962 \times 10^6}{300}\ \dfrac{kN}{m^2} = 6.54\ MN/m^2$

35   When the mass of the loaded trolley is 200 kg,
gravitational force on trolley $= (200 \times 9.81)\ N = 1\,962\ N$
$\therefore$ normal force $= 1\,962\ N$
Since a force of 350 N is required to pull the trolley at constant speed, the friction force is 350 N.

Coefficient of friction $= \dfrac{friction\ force}{normal\ force}$

i.e. $\mu = \dfrac{350}{1\,962} = 0.178$

When the mass of the loaded trolley is increased, so that its total mass is 300 kg,
$$W = (300 \times 9.81)\ N = 2\,943\ N$$
$\therefore$ normal force $= N = 2\,943\ N$
friction force $= F = \mu N = (0.178 \times 2\,943)\ N = 524\ N$
Since the trolley is moving at uniform speed, $P = F$
$\therefore$ horizontal pulling force $= 524\ N$

Further self-assessment questions

36   While a ship is steering due north at a velocity of 20 km/hour, a helicopter is flying north west at a velocity of 150 km/hour. Determine the velocity of the helicopter as it would appear to an observer on the ship.

37   To an observer on a destroyer sailing due west at 35 km/hour, a cruiser appears to be sailing south west. If the actual speed of the cruiser is 30 km/hour, determine the two possible directions in which the cruiser could be moving.

38   An electric motor shaft accelerates from rest to 20 rev/s in 6 seconds. Assuming a uniform acceleration, find the angular acceleration in rad/s^2 of the shaft, and the number of revolutions made by the shaft when its speed reaches 20 rev/s.

39   A rotating shaft accelerates uniformly from rest and reaches a speed of 60 rev/min after a total of 300 revolutions. If this acceleration is maintained, determine:
(i)   the angular acceleration in rad/s^2,
(ii)   the time taken from rest for the shaft to reach a speed of 5 rev/s,
(iii)   the total number of revolutions made by the shaft in attaining a speed of 5 rev/s.

40   A car, travelling at 72 km/hour, is brought to rest under a constant deceleration in a time of 30 seconds. If the wheels of the car are 800 mm diameter, find:
(i)   the angular speed of the wheels, in rad/s, when the car is travelling at 72 km/hour,
(ii)   the angular deceleration of the wheels in rad/s^2.

41   A haulage drum of diameter 1 m rotates at 4 rev/s. A brake is applied, and the drum is brought to rest with uniform deceleration in a time of 30 seconds. Find:
(i)   the angular deceleration of the drum in rad/s^2,
(ii)   the length of cable wound onto the drum during the deceleration phase.

42   A conveyor belt accelerates in a uniform manner from rest to a velocity of 1.5 m/s. The belt is driven by pulleys of effective diameter 50 mm. Find:
(i)   the final angular velocity of the pulleys in rev/s,
(ii)   the angular acceleration of the pulleys in rad/s^2, if the time taken to reach a velocity of 1.5 m/s from rest is 2.5 seconds.

43   A car with wheels 800 mm diameter is uniformly accelerated from a speed of 36 km/hour to a speed of 54 km/hour in a time of 20 seconds. The car travels at this speed for a time of 60 seconds, when it is retarded uniformly at a rate of 1 m/s^2 until it is stationary. Find the time from the beginning to end of the journey, and the number of revolutions made by the wheels of the car.

44   A machine tool table is driven by a constant force of 360 N. The table has a mass of 1 800 kg, and reaches a speed of 0.1 m/s from rest in a uniform manner. Find the time taken to reach this speed.

45   A train of mass 100 tonnes is travelling at 80 km/hour. Find the resisting force required to stop the train with uniform deceleration in a time of 1 minute.

Solutions to self-assessment questions

36   The vector diagram is drawn in Figure 302. The velocity of the helicopter relative to the ship is 107 km/hour at an angle of $52°$ west of north.

37   The vector diagram is drawn in Figure 303. The two possible courses of the cruiser are $11°$ north of west or $79°$ north of west.

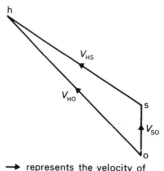

$\xrightarrow{}$
os   represents the velocity of the ship relative to a fixed point

$\xrightarrow{}$
oh   represents the velocity of the helicopter relative to a fixed point

Figure 302   *Solution to self-assessment question 36*

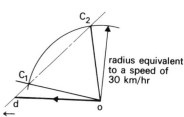

$\xleftarrow{}$
do represents the velocity of the destroyer. The vectors $\overleftarrow{oc}_1$ and $\overleftarrow{oc}_2$ represent the two possible courses of the cruiser

Figure 303   *Solution to self-assessment question 37*

38 $\alpha = 20.94$ rad/s^2, $\theta = 60$ revolutions.

39   (i) $\alpha = 0.0105$ rad/s^2
  (ii) $t = 2\,992$ s
  (iii) $\theta = 9\,276$ revolutions

40   (i) $\omega_1 = 50$ rad/s
  (ii) $\alpha = -1.667$ rad/s^2

41   (i) $\alpha = -0.838$ rad/s^2
  (ii)   length of cable = 188.5 m

42   (i) $\omega_2 = 9.55$ rev/s
  (ii) $\alpha = 24$ rad/s^2

43   Time for deceleration phase is 15 s
total time for journey = 95 s
Angular displacement of wheels in:
(i)   acceleration phase is 125 rad
(ii)   constant velocity phase is 2 250 rad
(iii) deceleration phase is 281.25 rad
Total number of revolutions = 422.7

46   A planing machine table, mass 1 200 kg, accelerates uniformly from rest to 0.2 m/s over the first 300 mm of its stroke. Calculate the force necessary to produce this acceleration.

47   A car of mass 1 tonne pulls a trailer of mass 0.5 tonne. The car and trailer accelerate uniformly from rest to a velocity of 40 km/hour in a time of 20 seconds. Determine the horizontal force acting on the towing pivot of the car, and the total force required to accelerate the car and trailer.

48   A mass of 400 kg is pulled along a rough horizontal floor with uniform speed by a horizontal force of 900 N. If an extra mass of 200 kg is placed on the first mass, calculate the horizontal force required to move the total load with uniform speed. Assume $g = 9.81$ m/s^2.

49   The moving table of a machine tool has a mass of 200 kg and slides on horizontal guides. The coefficient of friction between the table and the guides may be assumed constant at 0.1. If the table moves freely, find the retarding force on the table, and hence the time to bring it to rest from a speed of 0.9 m/s. Assume $g = 9.81$ m/s^2.

Solutions to self-assessment questions

44 $t = 0.5$ s

45   Resisting force = 37 kN

46   Accelerating force = 80 kN

47   Force on pivot = 278 N
Accelerating force = 834 N

48   Coefficient of friction = 0.229
Force required to move load = 1.35 kN

49   Retarding force = 196.2 N
Deceleration of table = 0.981 m/s^2
$t = 0.917$ s

Topic area: Heat

Section 1   Expansion

After reading the following material, the reader shall:

1   Describe expansion and solve simple problems.
1.1   Describe the relationship between temperature and increase of
   (*a*)   linear dimensions for solids,
   (*b*)   volumes for liquids.
1.2   Define the coefficients of expansion.
1.3   Solve simple problems involving changes of temperature and dimensions for solids and liquids.

The scientists following in the wake of intellectual giants such as Isaac Newton (1642-1727) and Gottfried von Leibniz (1646-1716) found great difficulty in explaining a phenomenon (heat) that could be produced by combustion, by friction, by the rays of the sun, by the hammering of metal into shape, or by rapidly reducing the volume of a fixed mass of gas in a cylinder.

There was a period — about 1800 — when the importance of careful observations was beginning to be recognized. James Prescott Joule (1818-89) who discovered experimentally the mathematical relationship between heat and mechanical work, has long been recognized as a pioneer in the art of scientific observation. Scientists noted that some of the more common effects of heat are:

(i)   to make bodies glow when heated strongly enough. Iron for instance first glows a dull red. At about $900°C$ (1173K) it becomes bright red hot, at about $1200°C$ it is yellow hot, whilst at temperatures above $1500°C$ the iron glows white hot.

(ii)   to cause bodies to change their state (or phase), e.g. to convert a solid into a liquid, or a liquid into a gas.

(iii)   to cause changes in the properties of a material, e.g. the electrical resistance of a conductor varies with temperature.

(iv)   to bring about chemical changes. This particular effect of heat has been of great importance in the development of those polymers called 'thermosetting'.

(v)   to cause a body to undergo expansion. The amount by which a body changes shape depends upon the original size, the change in temperature that occurs and the material of which the body is made.

Figure 304 diagrammatically shows these effects. The higher the temperature of a body, the more heat that body contains. However, it

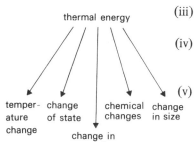

Figure 304   *Some effects of heat*

cannot be said that one body possesses more heat than another body just because its temperature is higher. A glass of hot water is plainly at a higher temperature than a bucket of cool water; yet the larger quantity of cool water will certainly melt more ice than the small volume of hot water. Despite its higher temperature, the hot water contains less thermal energy than the cool water.

The temperature of a body may be thought of as the degree of hotness relative to some chosen point (e.g. the melting point of ice, absolute zero etc.). Heat, however, can be envisaged as energy which is in the process of transfer between a body (or system) and its surroundings, the energy transfer being brought about by temperature differences.

Temperature and heat, then, mean different things, but the two concepts are closely interrelated. Most substances, whether gaseous, liquid or solid, expand when heated and contract when cooled. In engineering the expansion of materials can have both advantages and disadvantages. Accurate thermometers depend upon the regular expansion of the liquid metal mercury; but bridges with long spans must be designed to allow for a change in length between summer and winter. Experience helps designers to accommodate changes in size; an expansion bend would be incorporated in a steam pipe, the bend allowing movement to take place without damaging the pipeline.

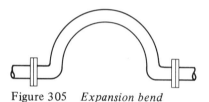

Figure 305   *Expansion bend*

When developing an understanding of expansion, it is necessary to consider the molecular structure of matter. All substances consist of molecules. In a solid (such as iron or copper) the molecules are packed fairly closely together. The molecules have mutual attraction for each other, and this helps to maintain the molecular pattern. When thermal energy is supplied, the molecules absorb the energy and vibrate at an increasing rate. As a result of this increased energy, the molecules are pushed a little farther apart, and thus, in a solid, they occupy a slightly greater volume. Expansion is caused by an increase in the spaces between the molecules.

In a liquid, the molecules move about much more freely; in fact, they can be thought of as being in random motion. They are not so constrained as the molecules in a solid, and in consequence liquids expand much more than solids when they are heated through the same temperature range.

A few solids have the property of hardly expanding at all when heated. This is because of their peculiar molecular structures. 'Pyrex', the glass used widely in cooking dishes and in laboratory equipment expands very little when heated, and in consequence the material does not crack as readily as normal glass. Quartz and fused silica have similar properties to Pyrex, and are used to make vessels which are required to maintain their shape and volume. Also, an alloy (of the metals nickel and steel) called 'Invar', expands by very small amounts when heated. Because of this most unusual property among metals, it finds many applications in such items as thermostats for gas cookers and in balance wheels in watches.

Most substances expand all the time that heat is being supplied, and contract when being cooled. Water, one of the most important substances on earth, is an exception to this rule. Over a small range of temperature it actually expands as it cools. As the temperature of water falls from 4°C to 0°C the liquid expands, and the resulting ice occupies a greater volume than did the water at 4°C. This is the reason why the solid ice floats on the parent liquid; ice is less dense than the water from which it is formed.

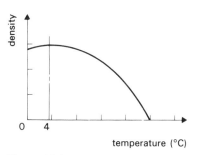

Figure 306 *The unusual expansion of water*

If water is contained in a pipe with the ends closed (as by a tap or valve), the large forces exerted as the water freezes to ice can crack the pipe. It is only when the temperature rises and the ice melts that the damage done to the pipe can be assessed. If there is a possibility of water freezing in a pipe in an industrial plant, then the plant designers have to ensure that all necessary precautions are taken.

In fact, the effects of expansions and contractions must be reckoned with constantly throughout industry. Expansion joints must be provided at each major section in building construction. In recent years European engineers have had to learn how to provide efficient expansion joints in oil-field piping. Designers of motor car engines have to allow for the different expansions of the pistons and the cylinders. Experience helps them to ensure that the lubrication does not break down, and that the pistons do not 'seize-up' inside the cylinders.

The forces exerted when materials are constrained and cannot change their dimensions have been put to use for many years. For example, wheels having separate metal tyres have been assembled by making the wheel slightly larger than the inside diameter of the tyre, and then heating the tyre until it can be placed in position round the wheel. When the tyre cools, it contracts and grips very firmly onto the wheel. In most cases no further means of securing the tyre is needed. Since the metals are sometimes made softer and less durable by the supply of this thermal energy, a more suitable method now employed uses solid carbon dioxide at about −80°C to cool the inner component. When the wheel returns to a normal temperature it tries to expand and pushes hard against the tyre.

The X shaped pieces of iron which can be seen on the outer walls of old buildings are the 'grip' plates through which iron tie rods pass. These rods are fitted when it is suspected that the walls are beginning to bulge outwards. The rods are heated along their length, and nuts tightened against the X-shaped plates whilst the tie rods are still hot. On cooling, the tie rods exert forces on the walls, preventing them from bulging further.

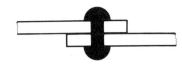

Figure 307   *Rivets*

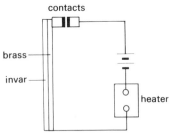

Figure 308   *Electrical thermostat*

Riveting is a method of joining plates together by a permanent form of clamping. The plates are usually made of metal, but plastic sheets are now frequently riveted together. One important application of riveting is the hot riveting of steel plates with steel rivets. This is still used in some engineering and shipbuilding activities. The rivets are heated to a red heat and then inserted through the hole in the plates. Lastly the end of the rivet is hammered into a mushroom shape. As the hot rivet cools, it tries to contract, making the joint between the plates more secure.

The 'bimetallic strip' makes use of the fact that different metals have different rates of expansion. If equal lengths of brass and 'Invar' are riveted together they can be used in the construction of devices such as thermostats. In the electrical thermostat shown, as the temperature rises the greater expansion of the brass ('invar' expands very little on heating) produces a curvature in the strip, and the electrical connection is broken. As the temperature falls below a pre-determined level the strip straightens, and the electrical circuit is again completed. By using the bimetallic strip the temperature in the heater can be kept nearly constant.

Another very common use of a bimetal strip is in the gauge in a motor car which shows the temperature of the cooling water. When the temperature of the cooling water deviates from normal the bimetal strip bends and rotates the pointer.

Open air tanks full to the brim of liquid will overflow as the temperature of the liquid rises — perhaps due to the rays of the sun. A toy balloon partially inflated in a cool room will expand and may well burst if it is held close to a fire. These two instances are simply examples of a general law that almost all substances expand on being heated.

Expansion is due principally to the fact that each molecule effectively occupies more space when the vigour of its vibrations is increased by the addition of thermal energy. Whatever may be the complete explanation of the ideas of expansion and contraction, it is important in engineering designs to be able to calculate accurately the size of a body at any temperature.

Expansion of a substance occurs in all directions simultaneously, but in engineering practice it is common to consider:

(i)   linear expansion which is applicable to such bodies as wires, cables and rods. These are bodies where a change in length is the factor of primary importance. Calculations of linear expansions refer only to solid bodies and do not refer to liquids or gases.

(ii) volumetric expansion, which is concerned with solids, liquids and gases. It is found experimentally that over a fair range of temperature, 1 m of aluminium expands by 0.000 024 metre for each degree Celsius that the temperature is raised. For copper the corresponding expansion is 0.000 017 metre, but the unusual alloy 'Invar' expands by only 0.000 002 metre. 'Pyrex' also expands little on heating; its expansion is 0.000 003 metre per metre of length for each degree Celsius rise.

The coefficient of linear expansion is defined as the change in length per unit of original length for each degree change in temperature. The coefficient of linear expansion is represented by the Greek letter α (alpha).

Hence if a rod has an initial length ℓ_1 and its temperature rises by one degree, the increase in length is $\alpha\ell_1$.

For a rise of t degrees, the increase in length is $\alpha\ell_1 t$.

Thus new length ℓ_2 = original length + increase in length
$$= \ell_1 + \ell_1 \alpha t$$
$$= \ell_1 (1 + \alpha t)$$

The way in which the length of the solid rod varies is as shown in Figure 309.

It has already been stated that one metre of copper expands by 0.000 017 metre when heated through a temperature rise of 1°C.

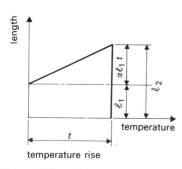

Figure 309 *Linear expansion*

α can then be written as $\dfrac{0.000\,017 \text{ metre}}{1 \text{ metre}} \times \dfrac{1}{1°C}$

Strictly, then, the units of α are $\dfrac{m}{m} \times \dfrac{1}{°C}$

However, the $\dfrac{m}{m}$ is usually cancelled, and the units of α are quoted as $\dfrac{1}{°C}$ or as per °C.

The following table gives some values of the coefficient of linear expansion for some common materials.

material	coefficient of linear expansion
aluminium	24×10^{-6} per °C
brass	19×10^{-6} per °C
copper	17×10^{-6} per °C
Invar	2×10^{-6} per °C
platinum	9×10^{-6} per °C
steel	11×10^{-6} per °C
glass (ordinary)	7 to 9×10^{-6} per °C
glass (Pyrex)	3×10^{-6} per °C

It may be noted from the above table that ordinary glass and platinum have approximately the same coefficient of linear expansion. In consequence it is possible to fuse platinum wires through a glass vessel without subsequent cracking occurring.

Example 1

Calculate the amount in millimetres by which a 5 metre length of aluminium will expand as its temperature is raised from 20°C to 40°C.

From the table, α for aluminium is 24×10^{-6} per °C.
Temperature rise $t = 40 - 20 = 20°C$.

$$
\begin{aligned}
\text{Increase in length} \quad &= \alpha \ell_1 t \\
&= 0.000\,024 \times 5 \times 20 \\
&= 0.002\,400 \text{ metre} \\
\text{Increase in length (mm)} \quad &= 0.002\,400 \times 1\,000 \\
&= 2.4 \text{ mm}
\end{aligned}
$$

Example 2

The length of a steel tape at 4°C is 30 metres. Calculate its length at 30°C. From the table, α for steel is 11×10^{-6} per °C.

$$
\begin{aligned}
\text{Temperature rise } t \quad &= 30 - 4 = 26°C \\
\text{Now } \ell_2 \quad &= \ell_1(1 + \alpha t) \\
&= 30(1 + 0.000\,011 \times 26) \\
&= 30(1 + 0.000\,286) \\
&= 30.008\,580 \\
\text{Length of tape at 30°C} \quad &= 30.008\,580 \text{ metre}
\end{aligned}
$$

Example 3

A shaft is to be a 'force fit' in the hub of a gear wheel. To assemble the wheel and shaft it is decided to cool the shaft so that it will pass freely through the hub of the wheel. At 15°C the steel shaft has a diameter of 100.00 mm. To what temperature must the shaft be cooled, if the designers decide that for ease of assembly the diameter must be reduced by 0.055 mm?

Change in diameter $\quad = 0.055$ mm
Now, change in diameter $= \ell_1 \alpha t$
where ℓ_1 is the original diameter of the shaft.

$$
\begin{aligned}
\text{From the table, } \alpha \quad &= 11 \times 10^{-6} \text{ per °C} \\
\text{Then, } 0.055 \quad &= 100 \times 11 \times 10^{-6} \times t \\
\text{where } t \quad &= \text{temperature fall during cooling} \\
\text{i.e. } t \quad &= \frac{0.055}{100 \times 11 \times 10^{-6}} \\
&= 50°C
\end{aligned}
$$

$$
\begin{aligned}
\text{Initial temperature} \quad &= 15°C \\
\text{Thus required temperature of shaft} \quad &= 15 - 50 \\
&= -35°C
\end{aligned}
$$

(Note that the final temperature could also be written as 238K.) Solid carbon dioxide could readily be used to produce this required temperature.

When dealing with the expansion of liquids, it is necessary to consider the increase in volume. The volume coefficient of expansion β (beta) is defined as the change in volume per unit of original volume for each degree change in temperature.

Let the original volume of a liquid be V_1 at a temperature t_1. Let the volume increase to V_2 when the temperature rises to t_2. Then, using similar ideas as were applied to linear problems,

$$V_2 = V_1(1 + \beta[t_2 - t_1])$$

Let $t_2 - t_1 = t$, the temperature change.

$$\text{Then } V_2 = V_1(1 + \beta t)$$

β for liquids is determined by experimental means. It is not constant over wide ranges of temperature, and average values are quoted in tables. In general, values of β are larger than values of α.

The table below gives the value of β at 20°C for a few common liquids.

ethyl alcohol	0.001 100 per °C
mercury	0.000 180 per °C
petroleum	0.000 900 per °C
water	0.000 207 per °C

Example 4

Calculate the reduction in volume when the temperature of 0.1 litre of ethyl alcohol is reduced from 35°C to -20°C. (Note that the temperature range is around the temperature 20°C, so the value given in the table will give a reasonably accurate answer.)

Reduction in volume $= V_1 \beta t$
where V_1 = initial volume
$\quad\quad \beta$ = volume coefficient of expansion
$\quad\quad t$ = temperature change

Thus reduction in volume = $0.1 \times 0.001\ 100 \times (35 - -20)$
$\quad\quad\quad\quad\quad\quad\quad\quad = 0.1 \times 0.001\ 100 \times 55$
$\quad\quad\quad\quad\quad\quad\quad\quad = 0.006\ 050$ litre

Self-assessment questions

1   Name three direct effects of supplying heat to a body.

2   Explain, preferably using sketches, why a bimetal strip (i.e. a strip consisting of two dissimilar metals riveted together) bends when heated.

3   Define the coefficient of linear expansion. Name the units that are normally used.

4   (*a*) Explain what is meant by the statement 'The coefficient of linear expansion of zinc is 0.000 026 per °C'.

    (*b*) A zinc rod is 1 metre long exactly when the temperature is 283 K. A copper rod at the same temperature is 1.002 300 metre long. When the two rods are both heated to 533 K, they are found to be just the same length. Calculate the coefficient of linear expansion of copper.

5   A metre scale, made of steel, is at its correct length when the temperature is 15°C. What will the length of the scale be when the temperature falls to 5°C? α for steel is 11×10^{-6} per °C.

6   Explain the statement 'The average volume coefficient of expansion of mercury is 0.000 180 per °C'.

Solutions to self-assessment questions

1   Heat (i) makes bodies glow, (ii) causes substances to change their state, (iii) causes the properties of a body to change, (iv) brings about chemical changes, (v) causes a body to expand.

These five effects are mentioned in the text. The reader may have suggested others; for instance, a difference in hotness may be used to generate a current in a thermocouple.

2   Suppose a strip of brass is riveted to a strip of steel. Brass expands almost twice as much as steel for a given increase in temperature (the table earlier in the text gave α for brass as 19×10^{-6} per °C and α for steel as 11×10^{-6} per °C). Consequently, when the compound strip is heated, it bends into an arc as shown in Figure 310.

3   The coefficient of linear expansion α is defined as the change in length per unit of original length for each degree change in temperature. α is strictly measured in metre/metre/°C, but is usually given as 'per °C'.

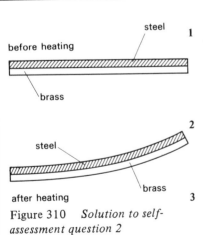

Figure 310   *Solution to self-assessment question 2*

Section 2
Specific and latent heat

After reading the following material, the reader shall:

2 Be familiar with specific heat capacity and specific latent heat.
2.1 Explain the significance of relationships between heat energy and temperature change.
2.2 Describe how changes of state occur without change in temperature.
2.3 Define specific heat capacity without reference to constant volume or pressure.
2.4 Define boiling and freezing points.
2.5 Define specific heat of fusion and specific latent heat of vaporization.
2.6 Solve simple problems involving the supply of thermal energy to solids and liquids.

After a body has been heated, it possesses energy which it did not have before. But the body on the whole has not altered its position in space, and thus has not changed its potential energy. Also, the body as a whole has acquired no motion, so it has not changed its kinetic energy. The conclusion which is reached by physicists is that the thermal energy must have been given to the separate molecules making up the body. It is molecular motion which is increased by added heat and decreased by the loss of heat.

Since heat is the kinetic energy of the molecules, heat is measured in the units of energy, i.e. in joules. The purpose of the work that follows is to consider how heat causes solids to melt into liquids, and causes liquids to vaporize.

Different materials require different amounts of heat to produce the same temperature rise. A galvanized iron roof in the direct rays of the sun may become unbearably hot to the touch, whilst a butt of water alongside it becomes only just warm.

Such everyday experiences show that some substances heat up much more readily than others. Water has a much greater ability to absorb heat than does iron. In fact water has the highest thermal capacity of any common substance. Because of this property, and also because it is cheap and available almost everywhere, water is the most universally used substance for carrying heat. Water is heated in boilers and pumped along pipes to 'radiators' where heat is transmitted to warm homes or factories or shops. Besides being used to carry heat to spaces where heat is required, water is used to carry heat away from spaces where it is not desirable. For instance, water circulates in metal jackets round the cylinders of motor car engines, picking up engine heat and thereby preventing excessive temperatures in the engine. The heated water is pumped to the car 'radiator', and there the unwanted heat is dissipated to the surrounding air.

More heat is required to raise a large mass of material through a given temperature range than is required for a small mass. To compare the thermal properties of different substances it is usual to quote the values of their specific heats. (The specific heat of a material is a measure of the heat required to raise unit mass of the material through one degree of temperature.)

Designers in industry frequently face problems which involve heat quantity calculations. Consider the following: an experimental electric furnace is rated at 4 kW and can be assumed to have an efficiency of 0.8. The furnace is to heat a 1 kg mass of copper to a temperature of $980^\circ C$. Assume that the original temperature of the copper is $20^\circ C$.

Solutions to self-assessment questions

4　(a)　This means that one metre of zinc expands by 0.000 026 metre when heated by $1^\circ C$.

(b)　Consider first the zinc rod.

$$\ell_2 = \ell_1 (1 + \alpha t)$$

ℓ_2 is the length at 533K

ℓ_1 is the length, 1 metre, at 283K

α is as given in the question.

i.e.　$\ell_2 = 1(1 + 0.000\,026 \times [533 - 283])$

$= 1(1 + 0.000\,026 \times 250)$

$= 1(1 + 0.006\,500)$

$= 1.006\,500$ metre

i.e. the zinc rod is 1.006 500 metre long at 533 K. However, the copper rod is known to be the same length at 533 K. Consider now the copper rod.

$$\ell_2 = \ell_1 (1 + \alpha t)$$

i.e.　$1.006\,500 = 1.002\,300\,(1 + \alpha \times 250)$

i.e.　$\dfrac{1.006\,500}{1.002\,300} = 1 + 250\alpha$

i.e.　$1.004\,190 = 1 + 250\alpha$

i.e.　$\alpha = \dfrac{0.004\,190}{250}$

$= 0.000\,017$

i.e. α for copper $= 0.000\,017$ per $^\circ C$

5　$\ell_2 = \ell_1 (1 + \alpha t)$

$= 1(1 + 0.000\,011 \times [5 - 15])$

$= 1(1 - 0.000\,110)$

$= 0.999\,890$ m

i.e. length of scale at $5^\circ C$ is 0.999 890 m

6　This statement means that one cubic metre of mercury would, on average, expand by 0.000 180 m^3 when heated by $1^\circ C$.

How many minutes will it take for the furnace to heat the copper to the required temperature?

Clearly, the designer of the furnace must take into account the materials which may be heated. *The quantity of heat needed to raise the temperature of 1 kg of a substance by 1 K (i.e. 1°C) is called the specific heat capacity of that substance.* The specific heat capacity of a substance is found by conducting careful experiments.

The units of specific heat capacity are joules per kilogram per degree. This is written, in SI, J/kg K. As one kelvin is equal to one degree celsius, it makes no difference to the calculations if the units of specific heat capacity are given as J/kg °C. The values in the table below are listed in both SI and celsius units.

substance	specific heat capacity (J/kg K)	specific heat capacity (J/kg °C)
lead	130	130
mercury	140	140
silver	235	235
copper	400	400
iron	440	440
aluminium	900	900
magnesium	1050	1050
turpentine	1800	1800
paraffin oil	2200	2200
ice	2100	2100
water	4200	4200

The specific heat capacity of a substance does not remain exactly constant, but varies somewhat with temperature.

The specific heat capacity is the heat required to heat one kilogram of a substance through one degree. It follows that if the specific heat capacity of a substance is c J/kg °C, then the heat required to raise the temperature of m kilograms of the material through t°C is mct joules.

Returning to the problem on the furnace, the value of the specific heat capacity of copper can now be used.

$$\text{temperature rise required} = 980 - 20 = 960°C$$
$$\text{heat energy required} = mct$$
$$= 1[\text{kg}] \times 400[\text{J/kg°C}] \times 960[°C]$$
$$= 384\,000 \text{ J}$$
$$\text{heat energy available/second} = 0.8 \times 4 \times 10^3$$
$$= 3\,200 \text{ J}$$
$$\therefore \text{time to achieve required temperature rise} = \frac{384\,000 \quad \text{J}}{3\,200 \quad \text{J/s}}$$
$$= 120 \text{ s}$$
$$= 2 \text{ minutes}$$

Example 5

Calculate the amount of heat needed to raise the temperature of 3 kg of aluminium by $100°C$.

$$\text{heat needed} = mct \text{ joules}$$

From the table,

$$c = 900 \text{ J/kg}°C$$

$$\therefore \text{ heat needed} = 3 \text{ [kg]} \times 900 \text{ [J/kg}°C] \times 100 \text{ [}°C]$$

$$= 270\,000 \text{ J}$$

Example 6

How many kilograms of water can be raised from 283 K to 303 K by the absorption of 840 kJ?

$$\text{from the table } c = 4\,200 \text{ J/kg K}$$

$$\text{increase in temperature } t = 303 - 283 = 20 \text{ K}$$

$$\text{heat absorbed} = 840 \text{ kJ}$$

$$= 840\,000 \text{ J}$$

$$\text{since heat absorbed} = mct$$

$$840\,000 = m \text{ [kg]} \times 4\,200 \text{ [J/kg K]} \times 20 \text{ [K]}$$

$$\text{i.e } m = 10 \text{ kg}$$

Example 7

A power station must disperse 4 200 MJ of waste heat energy every second when working at peak load. The most economical source of cooling water available is in a nearby river. Water at river temperature can be pumped through the condensers. After condensing the spent steam, the then heated cooling water is to be returned to the river at a point downstream from the point where the water was taken from the river.

From past records it is anticipated that the river will have a minimum flow rate of 0.5×10^6 kg/s. The cooling water extracted for cooling purposes must never exceed 25% of the river flow. The power station designers are informed by biologists that the fish life in the river can tolerate temperature rises that would be brought about by the cooling water being raised by $12°C$. Would the river be capable of providing all the cooling water required?

$$\begin{array}{ccccc}
\text{heat to be} & & \text{mass flow} & & \text{specific heat} & & \text{temperature} \\
\text{dispersed} & = & \text{per second} & \times & \text{capacity} & \times & \text{rise of} \\
\text{per second} & & & & \text{of water} & & \text{cooling water}
\end{array}$$

$$\text{i.e. } 4\,200 \times 10^6 = m \text{ [kg/s]} \times 4\,200 \text{ [J/kg}°C] \times 12 \text{ [}°C]$$

$$\text{i.e. } m = \frac{4\,200 \times 10^6}{4\,200 \times 12}$$

$$= \frac{10^6}{12} \text{ kg/s}$$

$$= 0.0833 \times 10^6 \text{ kg/s}$$

i.e. quantity of cooling water needed is 0.0833×10^6 kg/s

mass flow of river $= 0.5 \times 10^6$ kg/s

25% of this is available for cooling purposes

$$\text{mass flow of cooling water} = 0.25 \times 0.5 \times 10^6$$
$$= 0.125 \times 10^6 \text{ kg/s}$$

This means that the river is suitable to provide the cooling water to the power station.

Change of state

Gases can be changed into liquids, and liquids into solids if the temperature is reduced sufficiently. Many substances can exist as a gas, as a liquid or as a solid. These are considered to be different *states* of matter, rather than different kinds. The transition from one state to another is called a change of state.

There are many instances in industry where change-of-state processes are of importance. Most metals are mined in the form of ores; frequently the metal is present in the form of an oxide. The basic procedure in obtaining pure metals is to heat the ore to its melting temperature, and then remove the impurities. Melting (a change of state) is the first step in the refining of many metals.

Petroleum is a very complex mixture of chemical compounds called hydrocarbons. At the oil refinery, chemical engineers separate out one useful compound from another, i.e. they separate petrol, diesel fuel, fuel oil, asphalt and other compounds from the crude oil. These separations are effected in fractionating towers, where the crude oil is heated. The more volatile hydrocarbons (those which evaporate very quickly) undergo a change of state first, then the next most volatile, and so on. These vapours are condensed (changing state) and are drawn off at appropriate points in the fractionating towers.

Steam turbines use water as the heat-transfer medium. The water is cycled back and forth from the liquid state to the vapour state, and is used over and over again. In the boiler, water is changed in state into superheated steam. After doing its work on the turbine rotor, the now spent steam passes into the condenser, where its state is changed back to water. The water is then returned to the boiler, and the cycle continues.

Most domestic refrigerators and the majority of industrial refrigerators are based on the principles of the change of state.

The process of refrigeration consists of removing heat from some space or body (where the heat is not wanted) and transferring it to some space or material where the presence of the heat is of no consequence. The medium for this heat removal is a vapour which is readily condensed into a liquid. The liquids used in refrigerators are called refrigerants. All refrigerants have the property of boiling at low temperatures, and the heat that boils them (causing a change in their state) flows from the space or the food being cooled. The vapour is next compressed, then cooled until it changes state to a liquid, when the cycle is ready to commence again.

These few examples drawn from the metallurgical, mechanical-power and refrigeration industries help to illustrate the extremely wide applications of change-of-state phenomena in the engineering industries.

To explore in more detail the processes of melting and boiling, consider what happens when ice below 0°C (i.e. below 273 K) is heated at a constant rate. The ice will warm up, then turn into water, which in turn will warm up and eventually turn into steam. This steam can be heated up if it is contained in a vessel and a heat supply maintained.

Consider these events in turn. Starting with a dish of chipped ice at a temperature below 0°C, the supply of heat will cause a rise in temperature. This rise will continue steadily until the temperature 0°C is reached. Then, although the heat supply is kept steady, the temperature rise ceases as the ice begins to melt. If the ice-water mixture is thoroughly mixed, no temperature change occurs until all the ice has disappeared.

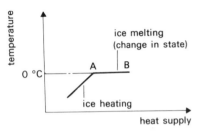

Figure 311   *Melting of ice*

When a solid (such as ice) is heated, the molecules vibrate more vigorously, but their average position remains unchanged. During a change of state such as solid to liquid (i.e. during melting) the temperature remains constant. The heat energy absorbed during this period causes the molecular vibrations to become so furious that the molecules break away from their average positions and move about much more freely. When a substance is solid, the energy of the molecules is mostly vibrational. When the solid melts, the molecules of the liquid are moving rapidly, and their energy is mostly kinetic.

Maintaining the constant supply of heat next causes the water to rise in temperature until a temperature of 100°C is reached. Provided that the air above the water is at normal atmospheric pressure, bubbles of vapour will form all through the water, and the liquid is said to be boiling. The temperature will remain constant at 100°C until all the water has been turned into vapour.

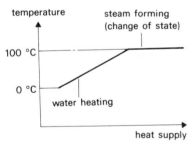

Figure 312   *Steam formation*

On heating a liquid, the kinetic energy of the molecules is increased. The liquid turns into a vapour when the energy of the molecules is so great that all the bonds between the molecules are broken. Then the molecules rush about at great speeds in a completely random fashion.

If the water vapour (steam) is heated in a vessel, its temperature will rise. There is no subsequent change of state. Steam heated in this way is called superheated steam; it is the kind of steam used in the turbines in power stations.

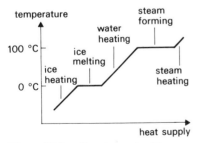

Figure 313   *Ice-steam stages*

The series of processes which produce superheated steam from the original ice is shown in Figure 313.

If the changes of state are reversed, then heat is given out instead of being absorbed. The condensation of steam to water results in the liberation of large amounts of heat, and the freezing of water also releases heat.

The changes of state in a number of other substances follow a similar pattern to that described for ice-water-steam.

Melting, freezing, evaporation and condensation
Materials can exist in three physical states — solid, liquid and gaseous. The changes of state generally occur in one of the following ways:

(i) Melting (or fusion) as a solid becomes a liquid.
(ii) Freezing as a liquid becomes a solid.
(iii) Evaporation (or boiling) as a liquid becomes a vapour.
(iv) Condensation as a vapour becomes a liquid.

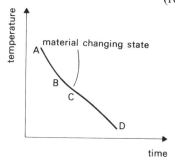

Figure 314   *Cooling curve for a non-crystalline substance*

In general, non-crystalline substances (such as glass) or amorphous substances (like the fats) do not have a definite melting point. The graph of temperature against time for the cooling of a sample of these materials is as shown in Figure 314. Although the rate of cooling is comparatively slow along the portion BC of the curve, at no time does the temperature remain constant, and there is no single temperature which can be called the melting (or freezing) point of the liquid. In non-crystalline and amorphous substances, properties such as viscosity, density, and electrical resistance change gradually whilst the change in state is occurring.

On the other hand, crystalline materials (like the metals) melt at definite temperatures. The melting point of these materials is the same temperature as the freezing point. If a crystalline material is heated to such a temperature that it is completely molten, and is then allowed to cool, the temperature-time graph is usually of the form shown in Figure 315. The exact melting points and the heat required to cause fusion of crystalline materials are factors of great importance in the metallurigical industries. Such properties as elasticity, hardness, toughness and malleability are determined by the structure of the finished metal, and the arrangement of the molecular pattern is 'tailored' by metallurgists using their knowledge of the effects of heat on the particular metal.

In a crystalline material the molecules are arranged in an orderly fashion. When the substance melts, this order is destroyed. As a result, many properties such as electrical resistance and specific heat undergo a sudden change at the melting point.

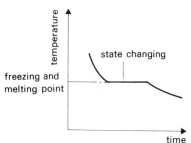

Figure 315   *Cooling curve for a crystalline substance*

A molecule situated deep in the body of a liquid is surrounded on all sides by other molecules, and the forces exerted on the molecule must nearly balance out. A fast moving molecule near the liquid surface, however, does not have balanced forces acting on it, and may escape into the atmosphere. The gradual loss of molecules in this fashion is known as evaporation. Since it is the faster moving molecules which leave the liquid, the average kinetic energy of the remaining molecules is less than the average kinetic energy of the molecules in the original liquid. The temperature of the remaining liquid decreases during evaporation, since a decrease in molecular motion means a decrease in thermal energy. A liquid such as petrol evaporates rapidly when exposed to the air; the rapid decrease in the heat content of the remaining liquid is the reason why such liquids feel cold when poured onto the hand.

If heat is supplied to a liquid, then the rate of evaporation increases. However, the molecules well below the liquid surface cannot escape, and so they evaporate internally, producing bubbles of vapour. The formation of these bubbles is only possible when the pressure created by the vapour is at least equal to the pressure acting on the surface of the liquid. The vapour bubbles are pushed up to the surface by the surrounding liquid and increase in size as they near the surface. The pressure in the liquid near the surface is less than the pressure at lower levels. This process is called *boiling*, i.e. boiling is rapid evaporation from all parts of a liquid. Boiling takes place at a particular temperature, this temperature depending only on the pressure which is exerted on the liquid. Boiling is different from evaporation in that evaporation takes place at all temperatures, and affects only molecules which are in the vicinity of the surface.

Boiling point is defined as that temperature at which the vapour pressure within the liquid is equal to the pressure acting on the surface of the liquid.

When a substance absorbs or rejects heat whilst remaining in one particular state (e.g. stays as a solid), then the thermal energy interchange causes a change in temperature.

Sensible heat is that heat which causes a change in temperature. The word sensible is used since the temperature change can be noted by the senses.

On Figure 311 there is no change in temperature between points A and B; there the solid may be melting to become a liquid, or if heat is being rejected a liquid may be changing its state and becoming a solid. The heat absorbed or rejected during such a change is called the *latent heat of fusion* (the name latent, meaning hidden, was given to this heat many years ago when the phenomenon was not understood). This quantity of heat can be determined only by experimental means. For example, if 1 kg of ice at 0°C is melted into water at 0°C, the latent heat absorbed has been found to be 335 kJ. The specific latent heat of fusion of ice is thus 335 kJ/kg. (Note: The word 'specific' refers to 1 kg of a substance.)

Example 8

A particular aluminium alloy changes from solid to liquid at 660°C; the specific latent heat of fusion is 410 kJ/kg. The specific heat capacity of aluminium is 0.9 kJ/kg K.

How much energy, in kilojoules, is needed to melt a casting of mass 10 kg, the casting being at a temperature of 10°C before being fed into the furnace?

energy needed = energy to raise the metal up to the melting temperature + energy to convert the solid metal into liquid

energy to raise the metal from 10°C to 660°C
$$= \text{mass} \times \text{specific heat capacity} \times \text{temperature rise}$$
$$= 10 \text{ [kg]} \times 0.9 \text{ [kJ/kg K]} \times (660 - 10) \text{ [K]}$$
$$= 5\,850 \text{ kJ}$$

energy to melt 10 kg of metal
$$= \text{mass} \times \text{specific latent heat of fusion}$$
$$= 10 \text{ [kg]} \times 410 \text{ [kJ/kg]}$$
$$= 4\,100 \text{ kJ}$$

∴ energy needed = (5 850 + 4 100) kJ
= 9 950 kJ

During a change of state from liquid to vapour the heat energy performs two tasks: (i) it overcomes the forces which keep the molecules fairly close together, and (ii) it provides the new vapour with sufficient energy to push back the atmosphere surrounding the liquid and new vapour. To appreciate the considerable change in volume which occurs on the evaporation of water, it may be noted that the volume of steam produced (at standard temperature and pressure) is almost 1 700 times as great as the volume of water from which it was formed.

The specific latent heat of vaporization of a substance is the amount of heat needed to change 1 kg of the substance from liquid to vapour, the temperature remaining constant.

For example, at standard atmospheric pressure 2 250 kJ are needed to turn 1 kg of water at 100°C into steam at the same temperature and pressure. This is a very large amount of energy, and it gives an idea of why a scald caused by steam is so damaging. When a person comes into contact with steam, this condenses, giving out large amounts of energy. This release of energy causes scalding.

Latent heats, like specific heats, are not constant. Their variations with temperature are related to the way in which energy is distributed among the atoms making up a material. Only since the quantum theory was put forward by Max Planck in 1900 has this subject been adequately understood. The quantum theory deals with the behaviour of matter on the atomic scale.

Example 9

(i) Find the amount of energy in joules needed to raise the temperature of 5 kg of water from 50°C to 100°C. Specific heat capacity of water is 4 200 J/kg K.

(ii) Find the additional energy needed to convert the water at 100°C into steam at 100°C. Assume that all the heating takes place at standard atmospheric pressure. Specific latent heat of vaporization of water is 2 250 kJ/kg.

(i) Heat needed to raise water temperature from 50° to 100°C	= mass × specific heat capacity × temperature rise
	= 5 [kg] × 4 200 [J/kg K] × (100 − 50) [K]
	= 1 050 000 J
	= 1.05×10^6 J
(ii) Heat energy needed to convert 5 kg of water into steam	= mass × specific latent heat of vaporization
	= $5 \times 2 250 \times 10^3$
	= $11 250 \times 10^3$
	= 11.25×10^6 J

Self-assessment questions

7 (a) Define 'specific latent heat of fusion'.

(b) Using the vertical axis to represent temperature and the horizontal axis to represent time, sketch the graph that would be obtained for the crystalline metal zinc as it cools slowly from the liquid state to the solid state. On the graph identify:

(i) the points where solidification starts and finishes,

(ii) the part of the graph associated with sensible heat,

(iii) the part of the graph associated with latent heat of fusion.

8 Write explanatory notes on the following:

(a) A large quantity of water placed in an unheated greenhouse will help to prevent frost from damaging the plants.

(b) A steam burn is usually much worse than a burn caused by hot water.

(c) A sponge pudding is much less likely to burn the mouth than a jam pudding at the same temperature.

9 Select the most comprehensive answer, 'Heat' is most closely related to:
   (*a*) chemical changes,
   (*b*) physical changes,
   (*c*) temperature,
   (*d*) energy.

10 If heat is supplied at a constant rate to a piece of ice, the temperature of the material first rises, then remains constant for some time, and afterwards begins to rise again. Sketch a graph, and indicate on it what happens during the three sections. What becomes of the heat supplied during the time the temperature remains constant?

11 Complete the following statements:
   (i) The amount of heat needed to change 1 kg of a substance from liquid to vapour at constant temperature is called _____.
   (ii) Sensible heat is the heat which causes _____.

12 Write explanatory notes on:
   (i) the evaporation of a pool of water, referring particularly to the kinetic energy of the molecules,
   (ii) how boiling differs from evaporation.

13 Define (*a*) boiling, (*b*) boiling point; in (*b*) make reference to the pressure acting on the surface of the liquid, and the vapour pressure inside the liquid.

14 Define the freezing point of a crystalline substance.

15 Why is it incorrect to talk of the 'freezing point' of a piece of glass?

16 Define (i) specific heat capacity, (ii) specific latent heat of vaporization.

17 Calculate the heat interchange needed in kJ to change:
   (*a*) 10 kg of ice at 0°C to water at 0°C.
   (*b*) 10 kg of steam at 100°C to water at 40°C.
   (*c*) 10 kg of ice at 0°C to water at 90°C.

   Specific latent heat of fusion of ice is 335 kJ/kg.
   Specific latent heat of vaporization of water at 100°C and standard atmospheric pressure is 2 250 kJ/kg.
   Specific heat capacity of water is 4 200 J/kg °C.

18 Complete the following statements:
   (i) In a solid, the energy of the molecules is mostly _____.
   (ii) In a liquid, the energy of the molecules is mostly _____.

19 As a liquid boils and becomes a vapour, the temperature remains constant. Name two tasks performed by the energy supplied during the evaporation of the liquid.

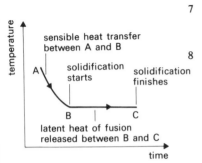

Figure 316   *Solution to self-assessment question 7*

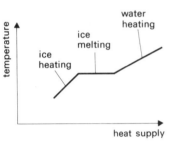

Figure 317   *Solution to self-assessment question 10*

Solutions to self-assessment questions

7   (*a*)   The specific latent heat of fusion is the amount of heat needed to change 1 kg of a substance from a solid to a liquid, the temperature remaining constant.

(*b*)   Figure 316 illustrates the points detailed.

8   (*a*)   The latent heat of fusion of water is high, and the large quantity of water will absorb much heat before the temperature in the greenhouse begins to fall.

(*b*)   When steam condenses on the skin of a person, it gives out its latent heat of evaporation. When hot water cools on contacting human skin, it gives out sensible heat. For equal masses of steam and water, the heat released by the condensing steam is much higher than that available from the cooling water. In consequence, a steam burn is likely to be much more serious.

(*c*)   Jams contain a high proportion of water. Since the specific heat capacity of water is higher than the specific heat capacity of the sponge pudding material, the sponge pudding is less likely to burn the mouth.

9   (*d*)   Heat is considered now to be the kinetic energy of the molecules in the material.

10   The sketch should look similar to Figure 317. During the period that the temperature remains constant, the heat energy supplied causes the molecular vibrations to become so vigorous that the molecules break away from their relatively fixed positions in the solid material and move about much more freely. The heat supplied has effectively converted the vibrational energy of the molecules into energy of motion; the ice has melted to water.

11   (i)   the specific latent heat of vaporisation.

(ii)   causes a change in the temperature of a body.

12   (i)   Fast moving molecules near the surface of the pool can escape into the atmosphere. They do not normally fall back into the pool. In consequence the water shrinks in volume. Over a period of time, all the molecules are likely to escape.

(ii)   Evaporation takes place at all temperatures, and affects only molecules which are close to the surface of the liquid. Boiling, on the other hand, takes place at a particular temperature, and it takes place throughout the liquid.

13   (*a*)   Boiling is rapid evaporation from all parts of a liquid. It occurs at a definite temperature, this temperature depending only upon the pressure exerted on the surface of the liquid.

(*b*)   Boiling point is defined as that temperature at which the vapour pressure within the liquid is equal to the pressure acting on the surface of the liquid.

14   The freezing point of a crystalline substance is that specific temperature at which the liquid rejects heat, and turns into a solid.

15   Glass is a non-crystalline substance, and observations show that it does not freeze (or melt) at one temperature. The freezing takes place over a range of temperature. Materials like glass, then, do not have a freezing point.

16   (i)   The quantity of heat needed to raise the temperature of 1 kg of a substance by 1 K (1°C) is called the specific heat capacity of that substance.

(ii)   The specific latent heat of vaporization of a substance is the amount of heat needed to change 1 kg of the substance from liquid to vapour, the temperature remaining constant.

17   (*a*)   In this case, the heat needed to produce a change of state is all that is required. The specific latent heat of fusion of ice is 335 kJ/kg. Thus to convert 10 kg of ice at 0°C into 10 kg of water at 0°C requires
10 [kg] × 335 [kJ/kg], i.e. 3 350 kJ.

20 The table below gives details of five substances, A, B, C, D, E. These substances are all of a crystalline nature when in the solid state.

substance	specific heat capacity (J/kg °C)	boiling point (°C)	melting point (°C)
A	2 200	60	− 100
B	4 200	100	0
C	140	361	− 40
D	2 500	300	16
E	385	−	1 100

(a) Which substances are most likely to be metals?
(b) Which substances are liquid at 65°C?
(c) Which substances are likely to be in vapour form at 70°C?

Solutions to self-assessment questions

(b) In this case heat must be extracted from the steam to make it condense to water at 100°C, and then further heat is removed to cool the water from 100°C to 40°C.

heat to condense steam = mass (kg) × specific latent heat of vaporization (kJ/kg)
= 10 × 2 250
= 22 500 kJ

heat needed to cool the water from 100°C to 40°C = mass × specific capacity × fall in temperature
= 10 [kg] × 4 200 [J/kg°C] × (100 − 40) [°C]
= 2 520 000 J
= 2 520 kJ

Thus total heat extracted = 22 500 + 2 520
= 25 020 kJ

(c) In this case the heat is needed firstly to melt the ice, and then to raise the resulting water at 0°C to 90°C.

heat to melt 10 kg of ice to water at 0°C = mass × specific latent heat of fusion
= 10 [kg] × 335 [kJ/kg °C]
= 3 350 kJ

heat to raise 10 kg of water from 0°C to 90°C = mass × specific heat capacity of water × temperature rise
= 10 [kg] × 4 200 [J/kg °C] × (90 − 0) [°C]
= 3 870 000 J
= 3 780 kJ

Thus total heat needed = 3 350 + 3 780
= 7 130 kJ

18 (i) vibrational, (ii) kinetic.

19 As the substance changes from a liquid into a vapour, the heat energy (i) overcomes the forces which in the liquid keep the molecules fairly close together, and (ii) it provides the new vapour with enough energy to push back the atmosphere surrounding the liquid and new vapour.

Further self-assessment questions

21   Calculate in kilojoules the heat that must be extracted from 10 kg of steam at 100°C to eventually produce 10 kg of ice at −20°C.
Specific heat capacity of ice is 2 100 J/kg °C.
Specific latent heat of fusion of ice is 335 kJ/kg.
Specific latent heat of vaporization of water is 2 250 kJ/kg.
Specific heat capacity of water is 4 200 J/kg °C.

22   A body of mass 2 kg is raised in temperature through 180°C as it absorbs 180 kJ. Calculate the specific heat capacity of the body.

23   In a cooling system 44 kg of oil per hour are cooled from 140°C to 45°C by means of circulating water. The temperature of this water must not rise by more than 5°C. How much cooling water (in kg per hour) is needed? The specific heat capacity of the oil is 2 100 J/kg °C. Specific heat capacity of water is 4 200 J/kg °C.

24   A bearing absorbs 12 kW in overcoming frictional resistance. (i) How much heat per minute (in kilojoules) will this frictional resistance produce? (ii) The bearing is cooled by passing oil through it. The flow of oil is kept constant at 48 kg per minute. The designers wish the oil to remove 75% of the heat produced by friction. Assuming the specific heat capacity of the oil to be 1 950 J/kg °C, by how many degrees will the temperature of the oil rise as it passes through the bearing?

25   An iron tyre has an internal diameter of 0.5 m when at 288 K. In order to be shrunk onto a wheel, the tyre is heated so that the diameter increases to 0.5045 metre. If the coefficient of linear expansion of the iron is 0.000 012 per °C, to what temperature must the tyre be heated?

26   A platinum wire is 0.2 m long when in ice at 0°C. When placed in steam at 100°C it extends by 0.18 mm. Calculate the coefficient of linear expansion of the platinum.

27   Steel cables on the suspension section of the San Francisco Bay Bridge are 3 000 m long on the coldest day of the winter when the temperature falls to −4°C. If on the hottest summer day the temperature rises to 31°C, what is the maximum annual variation in the cable length? (Take the value of α from the table on p. 235.)

Solution to self-assessment question

20   (*a*)   Metals, which heat up readily, have low specific heat capacities. Thus C and E are most likely to be metals. C is the metal mercury which melts at a low temperature. This makes it very useful in many industrial applications.

    (*b*)   B, C, D.

    (*c*)   A.

28 The volume of mercury in a thermometer is 3 000 mm^3 when the temperature is 10°C. What will be the volume of the mercury when the temperature is raised to 110°C. (Take the value of β from the table on p. 237.)

Solutions to self-assessment questions

21
Heat to condense steam	=	22 500 kJ
Heat to cool water to 0°C	=	4 200 kJ
Heat to condense water	=	3 350 kJ
Heat to cool ice to −20°C	=	420 kJ
Total	=	30 470 kJ

22   500 J/kg °C.

23   418 kg per hour.

24   (i)   720 kJ per minute.
(ii)   5.8°C.

25   1 038 K.

26   0.000 009 per °C.

27   1.155 m.

28   3 054 mm³.

Index